5 STEPS TO A 5™

AP Environmental Science

2016

AP Environmental Science

2016

Linda D. Williams

New York Chicago San Francisco Athens London Madrid Mexico City
Milan New Delhi Singapore Sydney Toronto

1 2 3 4 5 6 7 8 9 0 RHR/RHR 1 2 1 0 9 8 7 6 5

ISBN 978-0-07-184625-7
MHID 0-07-184625-5
ISSN 2151-2787

e-ISBN 978-0-07-184626-4
e-MHID 0-07-184626-3

The series editor was Grace Freedson, and the project editor was Del Franz.
Series design by Jane Tenenbaum.

ABOUT THE AUTHOR

LINDA D. WILLIAMS is a nonfiction writer with specialties in environmental science, medicine, and space and is trained as a microbiologist. Her work has ranged from biochemistry and microbiology to genetics and human enzyme research. During her career, she has worked as a lead scientist and technical writer for NASA and McDonnell Douglas Space Systems, and served as a science speaker for the Medical Sciences Division at NASA–Johnson Space Center. Ms. Williams has developed collaborative projects with researchers at the Weiss School of Natural Sciences at Rice University, Houston, Texas, and the University of Arkansas for the Medical Sciences. Her ability to make scientific subjects easily understandable has translated into several science books for all ages, including four books in the *McGraw-Hill Demystified Series* (i.e., *Chemistry, Earth Science, Environmental Science,* and *Nanotechnology Demystified*). For other books by Williams, visit her website, www.ldwilliams.com.

CONTENTS

PREFACE

Congratulations! You've arrived at the AP Environmental Science 5-Step Program. Your initiative to keep up with ever-changing environmental advancements and issues says you want to focus your attention on doing well in college-level biology, chemistry, ecology, chemical engineering, forestry, marine science, petroleum engineering, and agriculture classes among others. Environmental science is very much a multidisciplinary science.

The AP Environmental Science Exam, though extensive, is worth studying for and taking for college credit as well as to augment knowledge gaps in related subjects such as earth science.

This book can be used in addition to your regular environmental science book. Three different study programs are presented to match your available time before the test and go with different studying styles.

Now get started, follow the 5 Steps, and take the diagnostic and practice exams. Your scores will give you an idea of your strengths and weak spots. The Rapid Reviews and sample questions throughout will further familiarize you with what to expect on exam day. Good Luck!

ACKNOWLEDGMENTS

Illustrations in this book were generated with CorelDRAW®, Microsoft PowerPoint®, and Microsoft Visio® courtesy of the Corel and Microsoft Corporations, respectively.

National Oceanographic and Atmospheric Administration (NOAA), Environmental Protection Agency (EPA), and United States Geological Survey (USGS) statistics and forecasts were used where indicated.

This book is dedicated to the courage and unflagging energy of all the scientists, volunteers, and policy makers working to preserve and restore our environmental heritage for generations to come.

INTRODUCTION: THE FIVE-STEP PROGRAM

The Basics

At this point, you are either in an AP Environmental Science class, planning on taking one, or already schooled in many biological and ecological sciences and believe you may be able to get AP college credit with extra focus on the environmental sciences. All good reasons to take the AP Environment Exam! Solid effort and review of the material and questions offered in the *5 Steps to a 5: AP Environmental Science* study guide will give you a leg up and help ensure your score is the best it can be with the time you have to study.

Organization of the Book

This book takes you on a journey of the five steps needed to get ready for the AP Exam in environmental science. Each step will provide skills and approaches helpful in achieving a perfect 5.

First, you will be introduced to the five-step plan described in the book. Chapter 1 gives some background on the AP Environmental Science Exam. Chapter 2 lays out three study plans, based on the amount of time you have before the test. Chapter 3 has a Diagnostic Exam to give you an idea of how much you know already. In Chapter 4, various tips and different types of AP Exam questions are presented.

Chapters 5 to 19 offer a broad overview of the information important in an AP Environmental Science course. Although this is meant as review, some information may be new to you depending on your previous classes, teachers' particular focus, and how up-to-date you are on global environmental concerns. A summary and keywords are presented at the beginning of each chapter. A Rapid Review, along with multiple-choice and free-response questions and answers are also included.

Following the informational chapters, a practice exam is provided with both question types (i.e., multiple-choice and free-response) found on the AP Environmental Science Exam. After going through the study guide, you will be able to test your knowledge level on questions similar to those asked on past AP Environmental Science Exams. Should you find areas where your memory or understanding fog, check your answers with the explanations and note where you need to refresh. Go back to the subject chapters and review the information.

The appendices provide additional resources for your exam preparation, including:

- Units and Conversion Factors
- Acronyms
- Bibliography
- Websites

The Five-Step Program

Step 1: Set Up Your Study Program

The first step contains an overview of the AP Environmental Science Exam and an outline of associated topics. It also offers a time line to help with your exam review planning.

- **Bicycle Plan:** An entire school year: September to May
- **Hot Rod Plan:** One semester: January to May
- **Rocket Plan:** Six weeks: Quick exam brush up

Step 2: Determine Your Test Readiness

This step offers a diagnostic exam for a "sneak peek" at your test readiness. Take your time with the questions and do your best. Remember, it's intended to help you see how much you already know and how much preparation you need before you take the real exam.

Step 3: Develop Strategies for Success

Here the different question types are described with strategies for doing your best on the multiple-choice and free-response questions. You'll learn how to read and analyze the questions and whether or not to guess. How to organize your responses on the free-response questions is also presented.

Step 4: Review the Knowledge You Need to Score High

This step presents the bulk of the environmental science information you'll need to know to score well on the revised AP Environmental Science Exam. Since teachers and classes in different geographical areas may focus on one topic more heavily than another, this material helps fill in your knowledge gaps. The topics covered in Chapters 5 to 19 parallel those presented in a year-long AP class.

Step 5: Build Your Test-Taking Confidence

The final step in AP Environmental Science preparation gives you a chance to take a practice exam with advice on how to avoid common mistakes. Practice exam questions are not from actual AP environmental science exams, but mirror test material and question format.

Introduction to the Graphics Used in this Book

To emphasize particular strategies, we use several icons throughout this book. An icon in the margin will alert you that you should pay particular attention to the accompanying text. We use these three icons:

This icon points out a very important concept or fact that you should not pass over.

This icon calls your attention to a strategy that you may want to try.

This icon indicates a tip that you might find useful.

Set Up Your Study Program

CHAPTER 1

How to Approach the AP Environmental Science Exam

IN THIS CHAPTER

Summary: By keeping in mind that the global environment is a natural and changing system, environmental science concepts begin to seem like common sense.

Key Ideas

- Focus on the how and why of environmental processes rather than exact quantities and measurements.
- If something doesn't make sense, keep going. The bottom line often provides the information to figure out the answer.
- To memorize specialized words, make flash cards with the word on one side and the definition on the other. Get together with others and quiz each other.
- If you really don't understand something, ask the teacher for help.
- Although most environmental science is straightforward, don't try to learn it all in one sitting. Just like the food web, many concepts are interrelated and build on each other.

Background of the Advanced Placement Program

The Advanced Placement Program (AP) is a combined effort of teachers and students from high schools, colleges, and universities. Started in 1955, the program has helped students prepare for careers in dozens of course disciplines. Teachers from all levels prepare each AP course to cover the information, assignments, and skills of a related college course. A complete list of College Board AP courses and exams is available at AP Central http://apcentral.collegeboard.com/apc/public/courses/index.html. The student-focused AP website can be found at (www.collegeboard.com/apstudents).

Who Writes the AP Environmental Science Exam?

The AP Course Description and AP Exam are designed by members of the AP Environmental Science Development Committee (i.e., environmental scientists and educators). The AP environmental science course approximates one semester of a college introductory environmental science course. It includes biology, geology, chemistry, environmental studies, Earth science, and geography. The AP Environmental Science course is designed to provide first-year college students with a more advanced study of environmental science topics and/or fulfill a basic laboratory science requirement.

AP courses and exams are reviewed and revised based on survey results from around 200 colleges and universities, the College Board, key educational and disciplinary organizations, and the AP Environmental Science Development Committee members.

Who Takes the AP Environmental Science Exam?

Students taking the AP exam are either taking or planning to take biology, geology, chemistry, or other related classes. The great thing about AP environmental science is that many other subjects merge when studying the environment. Nature contains many interconnected components and concepts. In fact, most environmental disciplines and jobs call for a general knowledge of all these areas before a candidate is considered.

If you are taking an environmental class now, this book offers further clarification for (1) topics lightly touched upon in class, (2) topics you are particularly interested in and would like to know more about, and (3) the latest in environmental findings and their impacts.

Topics and Grading on the AP Environmental Science Exam

When I was in school, grades were everything. I stressed over every point and was frustrated I couldn't remember an "easy" answer. A whiz at memorization, I preferred fill-in-the-blank questions to multiple choice, where I would overthink the answers. Since then, I have come to realize that understanding concepts far outweighs a score on any one test. If relaxed during testing, your retention and comfort level increase exponentially and you score higher. Preparing for the AP Environmental Science Exam ahead of time will help you sail through.

The AP Environmental Science Exam is graded in June by college faculty members and high school AP teachers at the AP Reading. Thousands of Readers take part and are coordinated by a Chief Reader in each AP subject. Teachers who want to compare their students' grades with national percentiles can attain that goal by serving as a Reader. Grade distribution charts are also available on the AP student website.

The AP Environmental Science Exam is three hours long and divided between the multiple-choice and free-response sections. The multiple-choice part (60% of the final grade) with 100 multiple-choice questions tests a student's environmental science knowledge and understanding. The free-response part focuses on using learned principles. Students must organize answers through reasoning and use of their analytical skills. Four free-response questions are included in this portion of the exam (40% of the final grade). They include a data-set question, a document-based question, and two synthesis and evaluation questions. Examples of these kinds of questions are provided in this book. Calculator use is not allowed on either portion of the test.

The AP Environmental Science Exam comes from a variety of important topics such as

- Earth Systems and Resources
- Living World
- Population
- Land and Water Use
- Energy Resources and Consumption
- Pollution
- Global Change

These seven topic areas are weighted fairly equally (10–15%), with a higher weight (25–30%) given to the hot topic of air, land, and water pollution along with its human health impacts.

Multiple-Choice Question Scoring

Multiple-choice questions are recorded on a scan sheet graded by a computer. The computer adds up the number of correct responses, and that number is your multiple-choice raw score.

Free-Response Scoring

As stated before, you are required to answer four free-response questions. Sample questions can be found at the ends of Chapters 5 to 19 and at AP Central.

The College Board grades each test in the following way:

AP GRADE	COMBINED SCORE	QUALIFICATION
5	Mid 80s to 150	Extremely well qualified
4	Mid 60s to mid 80s	Well qualified
3	High 40s to mid 60s	Qualified
2	High 20s to high 40s	Possibly qualified
1	0 to high 20s	No recommendation

A report with your grade (1 to 5) will be sent to you in July. When you take the AP Environmental Science Exam, you may specify the colleges to which your AP score should be sent. These colleges and your high school will receive the scores for each AP exam you take along with earlier exam grades from previous years. You may request earlier grades be withheld from the report.

The Night Before the Exam

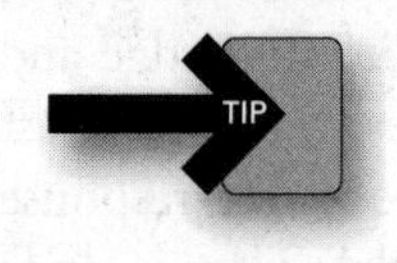

Last-minute cramming will just make you crazy or tired or both. The night before, look over troublesome topics (an hour or less), go do something fun and relaxing, and then get a good night's sleep.

The morning of the exam, wear your lucky T-shirt if you want, but dress comfortably. Eat a light meal (include some protein) before the exam. Allow plenty of time to get to the test site.

The Exam

Bring to the test location:

- A watch to keep track of your time
- Your school code
- Photo identification and social security number
- Two to three pencils and a nonsmudging eraser
- Black or blue pen for the free-response section
- Tissues, in case someone sitting near you has bathed in his or her favorite cologne or perfume and gives you a sneezing fit or a runny nose
- Determination to do your best!

Leave at home or in the car:

- Books, notes, flash cards, highlighters, rulers, etc.
- Music (CD players, iPods, etc., are not allowed)
- Cell phone, pager, or other electronic noise-making device

Planning Your Time

IN THIS CHAPTER

Summary: Just as every student is different, learning methods are different too. Depending on how you study and learn, choose a plan that matches the amount of time you generally need to absorb and understand new information.

Key Idea

✪ Pick the study plan that best matches your style of learning.

Preparing for the AP Environmental Science Exam

Depending on the amount of time until the test and the way you study best, the following study plans apply. You can tailor them to your personal learning style. Mix and match options depending on your previous knowledge of the topics.

Plan A: The Bicycle Plan (1 year)

- You are a long-range planner.
- You arrive at appointments quite early.
- You pay a lot of attention to detail.
- You believe in extreme preparation.
- You like being ready for anything.

Plan B: The Hot Rod Plan (1 semester)

- You expect challenges and plan changes.
- You prefer advance planning and can do without some small details.
- You like to be prepared, but welcome surprises.
- You know of a related course to take as added preparation.

Plan C: The Rocket Plan (6 weeks)

- You work best under pressure and short deadlines.
- You arrive at appointments just on time.
- You already know a lot about environmental science.
- You decided to take the exam six weeks before the exam.
- You are heroic and surprises energize you.

Table 2.1 gives a general timeline aligning the three study plans.

Table 2.1 General timeline aligning the three plans.

	PLAN A—BICYCLE PLAN	PLAN B—HOT ROD PLAN	PLAN C—ROCKET PLAN
Month	*(Full school year)*	*(1 semester)*	*(6 weeks)*
September–October	Introductory material, Diagnostic exam and Chapters 4–5		
November	Chapters 6–7		
December	Chapters 8–9		
January	Chapters 10–11	Chapters 5–7	
February	Chapters 12–13	Chapters 8–10	
March	Chapters 14–16	Chapters 11–14	
April	Chapters 17–19	Chapters 15–19	Skim Chapters 5–14; all Rapid Reviews; Practice exams
May	Overall review; Practice exams 1 and 2	Overall review; Diagnostic exam; Practice exams 1 and 2	Skim Chapters 15–19; Practice exams 1 and 2

If You Get Stuck

Throughout *5 Steps to a 5: AP Environmental Science*, I've used many examples and illustrations of environmental processes to help you visualize what is happening beneath, on, or above Earth's surface. In each chapter, there are keywords, a summary, multiple-choice questions, free-response questions, and a rapid-review section.

To refresh your memory or check certain details, look back at the chapter's information. Answer the questions; then, check how you did with the answers at the end of the chapter. You may want to linger in a chapter until you have a good handle on the material and get most of the answers right before moving on.

This book is divided into the seven major topic areas described in Chapter 1. When you have completed a chapter, go ahead and answer the questions. Take the chapter tests "closed book" when you are confident about your skills. Try not to look back at the text material when you are answering questions. The free-response questions are no more difficult than the multiple-choice questions, but serve as a more in-depth review with a timed period to complete your answers.

Along with the diagnostic exam in Chapter 3, there is a practice exam at the end of the book with questions like those in the individual chapters. Take the practice exam when you have finished all the chapter review questions and feel comfortable with the material as a whole.

You may want to have a friend or parent score your test without telling you the questions you missed. Then you won't be tempted to memorize answers to missed questions, but instead will go back and see if you got the point of the idea. After taking the practice exam, see how your score has improved from the diagnostic exam. This will confirm your strengths and highlight any areas where more study is needed.

Environmental science is not difficult, but does take some thought. Just plow through at a steady rate. If you are really interested in plate tectonics, for example, spend more time on Chapter 5. If you want to learn more about overfishing, allow more time in Chapter 12. You will complete your AP preparation according to your study timeline and have a solid knowledge base in environmental science.

Work Together

Often, when preparing for an AP exam, it's easier to stay on course when you have someone to bounce ideas off or encourage you when a topic seems especially complex. Think about finding someone who might want to get together every couple of weeks or monthly, depending on your study timeline, to review with you.

If You Have Questions, Ask!

If you come to a chapter that has you stumped, ask! If you're confused by atmospheric pollution specifics, ask a chemistry or ecology teacher. Likewise, if you're confused by a deforestation question, consult with a forestry, biology, botany, or ecology teacher.

Procrastination

One thing to keep in mind is that a study guide is only as good as the student who uses it. You won't have any advantage on the AP Environmental Science exam by putting off the review work. Pick a study timeline and stick to it. Good luck!

What's the Point?

This book brings current real-world environmental research and information to you in relation to its importance to the natural world and its inhabitants. Along the way, general concepts are summarized along with key ideas to remember for the AP exam and later academic training.

Remember, since environmental science is interdisciplinary, a broad array of topics from different scientific disciplines come together to explain environmental themes. Some of these include the following:

1. Science is a process.
   a. It is a method of learning more about the world.
   b. Information is gathered through careful testing and study.
   c. Conclusions are drawn after repeated testing and data analysis.
   d. Science and its discoveries change the way we understand the environment.

2. Energy conversions underlie all ecological processes.
   a. Energy is neither created nor destroyed but exists in different forms.
   b. As energy flows through systems, at each step more becomes bound and unavailable.
3. Earth is one huge interconnected system.
   a. Natural systems change over time and space.
   b. Geological time allows balancing of natural processes.
   c. Biogeochemical systems are able to recover from impacts to varying degrees.
4. Humans change natural systems.
   a. Humans have impacted the environment to a lesser or greater extent for thousands of years.
   b. Humans have increased their rate and level of impact on the environment through improved technology and population growth.
   c. Humans can modify energy production and use methods to be sustainable with environmental needs.
5. Environmental issues include a cultural and social context.
   a. To solve environmental problems, the role of cultural, social, and economic factors must be considered.
   b. Environmental issues must be approached in a global context.
6. Human survival depends on finding ways to reach sustainable systems and processes.
   a. A balanced combination of development and conservation is key.
   b. Managing natural and common regional and global resources is critical.
   c. Development of nonpolluting transportation fuels is crucial for economic as well as environmental reasons.

Determine Your Test Readiness

Take a Diagnostic Exam

IN THIS CHAPTER

Summary: A diagnostic exam allows you to figure out how much you know about the environment and what you might need to learn or review. After taking the exam, you can study what you missed. However, since science and engineering are always making new discoveries, looking over all the topics is a good idea. Additionally, the folks preparing the AP Environmental Science Exam may pull material from the latest environmental information.

Key Idea

- Connectedness of environmental science concepts, familiarization with exam topics, experience with multiple-choice questions, strength and weakness check, identification of topics to be reviewed.

Diagnostic Exam: AP Environmental Science

ANSWER SHEET

1 Ⓐ Ⓑ Ⓒ Ⓓ Ⓔ	26 Ⓐ Ⓑ Ⓒ Ⓓ Ⓔ	51 Ⓐ Ⓑ Ⓒ Ⓓ Ⓔ
2 Ⓐ Ⓑ Ⓒ Ⓓ Ⓔ	27 Ⓐ Ⓑ Ⓒ Ⓓ Ⓔ	52 Ⓐ Ⓑ Ⓒ Ⓓ Ⓔ
3 Ⓐ Ⓑ Ⓒ Ⓓ Ⓔ	28 Ⓐ Ⓑ Ⓒ Ⓓ Ⓔ	53 Ⓐ Ⓑ Ⓒ Ⓓ Ⓔ
4 Ⓐ Ⓑ Ⓒ Ⓓ Ⓔ	29 Ⓐ Ⓑ Ⓒ Ⓓ Ⓔ	54 Ⓐ Ⓑ Ⓒ Ⓓ Ⓔ
5 Ⓐ Ⓑ Ⓒ Ⓓ Ⓔ	30 Ⓐ Ⓑ Ⓒ Ⓓ Ⓔ	55 Ⓐ Ⓑ Ⓒ Ⓓ Ⓔ
6 Ⓐ Ⓑ Ⓒ Ⓓ Ⓔ	31 Ⓐ Ⓑ Ⓒ Ⓓ Ⓔ	56 Ⓐ Ⓑ Ⓒ Ⓓ Ⓔ
7 Ⓐ Ⓑ Ⓒ Ⓓ Ⓔ	32 Ⓐ Ⓑ Ⓒ Ⓓ Ⓔ	57 Ⓐ Ⓑ Ⓒ Ⓓ Ⓔ
8 Ⓐ Ⓑ Ⓒ Ⓓ Ⓔ	33 Ⓐ Ⓑ Ⓒ Ⓓ Ⓔ	58 Ⓐ Ⓑ Ⓒ Ⓓ Ⓔ
9 Ⓐ Ⓑ Ⓒ Ⓓ Ⓔ	34 Ⓐ Ⓑ Ⓒ Ⓓ Ⓔ	59 Ⓐ Ⓑ Ⓒ Ⓓ Ⓔ
10 Ⓐ Ⓑ Ⓒ Ⓓ Ⓔ	35 Ⓐ Ⓑ Ⓒ Ⓓ Ⓔ	60 Ⓐ Ⓑ Ⓒ Ⓓ Ⓔ
11 Ⓐ Ⓑ Ⓒ Ⓓ Ⓔ	36 Ⓐ Ⓑ Ⓒ Ⓓ Ⓔ	
12 Ⓐ Ⓑ Ⓒ Ⓓ Ⓔ	37 Ⓐ Ⓑ Ⓒ Ⓓ Ⓔ	
13 Ⓐ Ⓑ Ⓒ Ⓓ Ⓔ	38 Ⓐ Ⓑ Ⓒ Ⓓ Ⓔ	
14 Ⓐ Ⓑ Ⓒ Ⓓ Ⓔ	39 Ⓐ Ⓑ Ⓒ Ⓓ Ⓔ	
15 Ⓐ Ⓑ Ⓒ Ⓓ Ⓔ	40 Ⓐ Ⓑ Ⓒ Ⓓ Ⓔ	
16 Ⓐ Ⓑ Ⓒ Ⓓ Ⓔ	41 Ⓐ Ⓑ Ⓒ Ⓓ Ⓔ	
17 Ⓐ Ⓑ Ⓒ Ⓓ Ⓔ	42 Ⓐ Ⓑ Ⓒ Ⓓ Ⓔ	
18 Ⓐ Ⓑ Ⓒ Ⓓ Ⓔ	43 Ⓐ Ⓑ Ⓒ Ⓓ Ⓔ	
19 Ⓐ Ⓑ Ⓒ Ⓓ Ⓔ	44 Ⓐ Ⓑ Ⓒ Ⓓ Ⓔ	
20 Ⓐ Ⓑ Ⓒ Ⓓ Ⓔ	45 Ⓐ Ⓑ Ⓒ Ⓓ Ⓔ	
21 Ⓐ Ⓑ Ⓒ Ⓓ Ⓔ	46 Ⓐ Ⓑ Ⓒ Ⓓ Ⓔ	
22 Ⓐ Ⓑ Ⓒ Ⓓ Ⓔ	47 Ⓐ Ⓑ Ⓒ Ⓓ Ⓔ	
23 Ⓐ Ⓑ Ⓒ Ⓓ Ⓔ	48 Ⓐ Ⓑ Ⓒ Ⓓ Ⓔ	
24 Ⓐ Ⓑ Ⓒ Ⓓ Ⓔ	49 Ⓐ Ⓑ Ⓒ Ⓓ Ⓔ	
25 Ⓐ Ⓑ Ⓒ Ⓓ Ⓔ	50 Ⓐ Ⓑ Ⓒ Ⓓ Ⓔ	

Diagnostic Exam: AP Environmental Science

Time–45 minutes

Questions: 60

This exam has multiple-choice questions to help you gauge your strengths and weaknesses. Record your answers on the answer sheet provided. When you finish the exam, check your answers with the related material in the different chapters or topics of the book. Remember, it's not as important to get all answers correct as to spot your weak areas. If you're stuck on a question, review the material before going on. It will boost your confidence and help you retain it. Good luck!

Chapter 5

1. Primary waves (or P waves) got their names because they are

(A) poorly understood
(B) the fastest seismic wave type
(C) waves that arrive at seismic monitoring stations last
(D) the slowest seismic wave type
(E) known as Rayleigh waves

2. In plate tectonics, colliding plates along a convergent margin often form

(A) a ridge
(B) hotspots
(C) rifts
(D) a continental shelf
(E) an ocean trench

3. During volcanic explosions, more people are often killed by

(A) big rocks
(B) lava
(C) poisonous gases
(D) ash
(E) frightened mobs

4. The formation and movement of ocean and continental masses is known as

(A) mountain building
(B) plate tectonics
(C) subduction
(D) convection
(E) creep

Chapter 6

5. Relative humidity best describes

(A) wind currents
(B) point source humidity
(C) desertification
(D) the relationship between air temperature and water vapor
(E) a major drought factor

6. When wind changes speed or direction abruptly, it is known as

(A) gusty wind
(B) a front
(C) wind shear
(D) wind sock
(E) wind chill

7. The Earth's rotation is a major factor in

(A) the Coriolis effect
(B) gravitation
(C) the South Atlantic anomaly
(D) summer storms
(E) temperature

8. Which layer of the atmosphere is the most active with regard to weather?

(A) Biosphere
(B) Troposphere
(C) Ionosphere
(D) Stratosphere
(E) Thermosphere

Chapter 7

9. The zone of aeration is a soil layer where

(A) the spaces between particles are completely filled with water
(B) yearly tillage takes place
(C) nutrient-rich soil is located deep underground
(D) the spaces between particles are filled with air and water
(E) drought has caused cracks in the topsoil

10. When water enters an aquifer, it

(A) recharges
(B) becomes brackish
(C) evaporates
(D) is lost forever to human consumption
(E) loses ionization

11. Drainage basins with thousands of streams and rivers draining to them are also called

(A) calderas
(B) watersheds
(C) runoff zones
(D) cairns
(E) permeability basins

12. Water may spend as little as 2 days in the hydrologic cycle or as much as

(A) 1 month
(B) 1 year
(C) 100 years
(D) 1,000 years
(E) 10,000 years

Chapter 8

13. When rocks of one type are changed into another type by heat or pressure, they become

(A) more porous
(B) fine sediment
(C) metamorphic rock
(D) less compacted
(E) more opaque

14. Frost wedging represents what type of weathering?

(A) Saline
(B) Chemical
(C) Physical
(D) Solar
(E) Biological

15. Rocks are broken down into smaller pieces in all of the following ways except

(A) frost
(B) biological weathering
(C) unloading
(D) granular integration
(E) mechanical weathering

16. A chemical reaction with water that speeds up chemical weathering is called

(A) acid rain
(B) ozone
(C) sedimentation
(D) erosion
(E) basic rain

Chapter 9

17. Gause's principle explains that

(A) populations are independent of resource consumption
(B) no two species can fill the same niche at the same time
(C) organisms adapt and reproduce in a changing environment
(D) genetic changes are never random
(E) all living things are grouped into species or separate types

18. The close relationship between organisms of different species that may or may not help each other is called

(A) symbiosis
(B) predation
(C) realized niche
(D) founder effect
(E) macroevolution

19. Chance alterations in the genetic material of a cell's nucleus are known as

(A) genetic selection
(B) interspecific competition
(C) messenger RNA
(D) mitosis
(E) genetic drift

20. The measure of the number of different organisms, species, and ecosystems is called

(A) mutualism
(B) genetic drift
(C) bottleneck effect
(D) biodiversity
(E) adaptation

Chapter 10

21. A species' placement within a food chain or web is known as its

(A) niche
(B) trophic level
(C) predatory limit
(D) ecological pyramid
(E) biological cycle

22. The hydrologic cycle includes all the following except

(A) condensation
(B) evaporation
(C) respiration
(D) transpiration
(E) precipitation

23. Plants use sunlight in visible wavelengths to produce energy in a process called

(A) symbiosis
(B) concretion
(C) ionization
(D) photosynthesis
(E) halitosis

24. The main component of organic matter is

(A) carbon
(B) potassium
(C) calcium
(D) sodium
(E) iron

Chapter 11

25. What population dynamic principle is described by the equation $rN = \Delta N/\Delta t$?

(A) Exponential growth
(B) Life span
(C) Biotic potential
(D) Fecundity
(E) Urban sprawl

26. The rule of 70 is used by researchers to estimate a population's

(A) fertility
(B) survivorship
(C) use of resources
(D) doubling time
(E) density

27. What tool do scientists use to study and predict changes in a population?

(A) Biological cycles
(B) Age–structure diagrams
(C) Natality rates
(D) Emigration statistics
(E) Exponential growth

28. Logistic growth is obtained from the equation $rN(1 - N/K) = \Delta N/\Delta t$, where K is

(A) birth rate
(B) death rate
(C) carrying capacity
(D) a density-independent factor
(E) the environmental resistance factor

Chapter 12

29. Extracting DNA from one species and splicing it into the genetic material of another is known as

(A) mechanical engineering
(B) genetic engineering
(C) copyright infringement
(D) cell biology
(E) *in vitro* fertilization

30. All the following are pesticides except

(A) odonticides
(B) insecticides
(C) acaricides
(D) fungicides
(E) herbicides

31. Non-native (alien) species are transplanted by all the following, except

(A) ship's ballast water
(B) wind
(C) hitchhikers on another species
(D) humans (e.g., releasing exotic pets)
(E) swimming to a nearby area

32. With regard to plant produce, the term *GMO* stands for

(A) generally mild odor
(B) guaranteed multiyear organic
(C) genetically modified organism
(D) green malodorous onion
(E) genetically maintained organism

Chapter 13

33. Old-growth forests have

(A) low rainfall
(B) never been harvested
(C) limited species
(D) increased in recent years
(E) poor drainage

34. If a species can't adapt or find new habitat, it can become

(A) overgrown
(B) economically undesirable
(C) competitively aggressive
(D) extinct
(E) a foundation species

35. The large-scale, total elimination of trees from an area is known as

(A) crop control
(B) selective cutting
(C) green evolution
(D) wetlands
(E) deforestation

36. Deserts cover approximately what percentage of the Earth's land surface?

(A) 10%
(B) 25%
(C) 30%
(D) 45%
(E) 60%

Chapter 14

37. The area needed for a population to provide for itself and dispose of its waste is called its

(A) range
(B) ecological footprint
(C) naturalization zone
(D) realized niche
(E) sanitation cycle

38. The IUCN focuses the bulk of its environmental efforts on all of the following, except

(A) economic development of hot spots
(B) preserving natural processes
(C) protecting genetic diversity
(D) water purification
(E) upholding sustainability of wild species and ecosystems

39. How many toxic air pollutants are released yearly from U.S. mining operations as estimated by the Environmental Protection Agency?

(A) 50
(B) 100
(C) 200
(D) 275
(E) 400

40. Scientists use the IPAT equation ($I = P \times A \times T$) to find

(A) soil horizon depths
(B) a population's carrying capacity
(C) a population's impact on area resources
(D) beach erosion rates
(E) biodiversity of endangered species

Chapter 15

41. Which of the following does the United States depend on for 85% of its energy?

(A) Wind
(B) Geothermal
(C) Nuclear
(D) Fossil fuels
(E) Solar

42. The term for energy's rate of flow is

(A) meter/second
(B) joule
(C) gram
(D) power
(E) newton

43. Which fossil fuel needs little refining and is burned with 75–95% efficiency?

(A) Gasoline
(B) Tar
(C) Natural gas
(D) Diesel
(E) Jet fuel

44. When oil is distilled into gasoline, what percentage of the original energy is lost?

(A) 20%
(B) 25%
(C) 50%
(D) 60%
(E) 75%

Chapter 16

45. In 1898, which scientist coined the term *radioactivity?*

(A) Robert Rhodes
(B) Marie Curie
(C) Stephen Petkoff
(D) Alice Screws
(E) Richard Brannon

46. Uranium oxide, found in granite and other volcanic rocks, is also known as

(A) pitchblende
(B) talc
(C) tritium
(D) feldspar
(E) gneiss

47. Of the 439 nuclear power plants worldwide, how many are in the United States?

(A) 62
(B) 88
(C) 104
(D) 117
(E) 123

48. When nuclear core temperatures rise out of control, causing the core to melt through the Earth's crust, it is known as

(A) the Helsinki process
(B) a public policy concern
(C) fission
(D) INPO
(E) the China syndrome

Chapter 17

49. The Staebler-Wronski effect describes

(A) wind turbine wobble
(B) solar cell degradation
(C) isotope decay
(D) geothermal heat exchange
(E) hydroelectric flow rates

50. What additive is used in the United States to lower incomplete combustion emissions?

(A) Methanol
(B) Benzene
(C) Neon
(D) Ethanol
(E) Ammonia

51. Electricity sent out in small amounts to nearby points where it is needed is known as

(A) central power
(B) off-grid power
(C) green tags
(D) certified power
(E) distributed power

52. What percentage of United States power comes from geothermal energy?

(A) 5%
(B) 10%
(C) 15%
(D) 20%
(E) 30%

Chapter 18

53. Toxic pollutants trapped indoors and mixed with mold spores has led to

(A) fungi remediation
(B) decreased sick days
(C) sick building syndrome
(D) greater productivity
(E) construction fines

54. The Corporate Average Fuel Economy (CAFE) for motor vehicles was created by the

(A) Department of Energy
(B) Occupational Safety and Health Administration
(C) Clean Air Act
(D) Energy Policy and Conservation Act
(E) National Ambient Air Quality Standards

55. Hydrologists check the health of freshwater streams by measuring

(A) total organic carbon levels
(B) particulate levels
(C) dead zones
(D) sodium levels
(E) algal levels

56. Which of the following pH measurements from water could indicate acidic mine drainage?

(A) pH 5.1
(B) pH 6.8
(C) pH 7.3
(D) pH 8.0
(E) pH 8.2

Chapter 19

57. When atmospheric gases keep heat from being released back into space, allowing it to build up in the Earth's atmosphere, it is known as

(A) the El Niño–Southern Oscillation
(B) the greenhouse effect
(C) carbon sequestration
(D) the swamp gas outcome
(E) the smog effect

58. Which colorless, odorless, flammable hydrocarbon released by the breakdown of organic matter and coal carbonization is the second biggest additive to the greenhouse effect?

(A) Ozone
(B) Chlorophyll
(C) Methane
(D) Benzene
(E) Ammonia

59. Which atmospheric bodyguard has a crucial role in protecting life on Earth from radiation?

(A) Phosphorus
(B) Oxygen
(C) Nickel
(D) Helium
(E) Ozone

60. The difference in the climate's average state or its variability over an extended period of time is summarized as

(A) desertification
(B) the greenhouse effect
(C) glaciation
(D) climate change
(E) the El Niño–Southern Oscillation

› Answers and Explanations

Chapter 5

1. B—Primary waves travel 14 times faster than sound waves through air.

2. E—An ocean trench is formed when two converging plates collide.

3. C—Hot poisonous gases make breathing impossible, and people succumb quickly.

4. B—This is the umbrella theory geologists use to explain the Earth's activity.

Chapter 6

5. D—Relative humidity is the air's ability to hold moisture. Hotter air holds more moisture.

6. C—Wind shear is a factor in hurricane development and strength.

7. A—The Coriolis effect is not a function of gravity, but of the Earth's rotation and air movement.

8. B—The troposphere is the atmospheric layer where most weather takes place.

Chapter 7

9. D—The zone of aeration is closest to the soil's surface where spaces between soil particles contain both air and water.

10. A—Aquifers can be recharged through snow melt and infiltration.

11. B—The boundary between two watersheds is called a divide.

12. E—Residence time in groundwater can be many thousands of years.

Chapter 8

13. C—Metamorphic rock becomes denser and more compact when exposed to high heat.

14. C—Physical weathering breaks rock by mechanical means (e.g., splitting, smashing).

15. D—Rocks are broken by granular disintegration.

16. A—Acid rain happens through a process called dissolution, which dissolves minerals.

Chapter 9

17. B—Gause's principle explains that no two species can fill the same niche at the same time.

18. A—Symbiosis has three main types: mutualism, commensalisms, and parasitism.

19. E—Random protein breaks or other genetic changes are known as genetic drift.

20. D—Biodiversity includes all of the various types of organisms, species, and ecosystems.

Chapter 10

21. B—Trophic levels include producers, consumers, and decomposers.

22. C—Respiration is not a main component of the hydrologic cycle.

23. D—Photosynthesis is a plant energy process using visible light wavelengths.

24. A—Carbon is the primary component of all organic matter.

Chapter 11

25. C—Biotic potential is the rate of growth (r) times the number of individuals (N), which is equal to the change in the number of individuals (ΔN) in a population over a change in time (Δt). The r component describes an individual's average role in population growth.

26. D—Doubling time (years) is obtained by dividing 70 by the annual percentage growth rate.

27. B—Age–structure diagrams help scientists predict trends in a specific population.

28. C—Change in population number over time ($\Delta N/\Delta t$) is equal to exponential growth over time ($r \times N$) times the carrying capacity (K) of the population size (N). The term ($1 - N/K$) stands for the relationship between N, a point in time, and K, the number of individuals supported by the environment.

Chapter 12

29. B—Genetic engineering is often done to improve the quality and amount of plant yield.

30. A—Insecticides (kill insects), herbicides (kill plants/weeds), acaricides (kill spiders and ticks), and fungicides (kill molds and fungus) are types of pesticides.

31. E—Alien species are usually widely separated and not able to move easily to a new area.

32. C—Genetically modified organism.

Chapter 13

33. B—Old-growth forests have wide diversity and have never been harvested.

34. D—When a species can't adapt and every individual dies, the species becomes extinct.

35. E—Deforestation occurs when all trees are cut down and not replanted.

36. C—One third of the Earth is covered by deserts.

Chapter 14

37. B—Americans' ecological footprint (10 hectares/person) is larger than that of other countries.

38. A—The IUCN does not focus on the economic development of hot spots.

39. B—The EPA lists over 100 toxic air pollutants from U.S. mining operations yearly.

40. C—The equation solves for I (total impact) by using P (population size), A (affluence), and T (technology level) to project a population's influence on its local and regional resources.

Chapter 15

41. D—The United States has become increasingly dependent on fossil fuels for energy.

42. D—Power is the term used to describe the rate of the flow of energy.

43. C—Natural gas burns fairly cleanly and is considered a clean fuel.

44. E—The majority of oil's original energy is lost during distillation.

Chapter 16

45. B—French scientist Marie Curie coined the term *radioactivity* and, along with her husband Pierre, received the Nobel Prize for her work in radioactivity.

46. A—Pitchblende is a common name for uranium oxide ore.

47. C—Nearly 25% of the nuclear power plants worldwide are in the United States.

48. E—It was feared the core would melt through rock all the way to China.

Chapter 17

49. B—Thin-film solar cells of amorphous silicon experience a 15–30% decreased yield when exposed to sunlight.

50. D—Pure ethanol increases performance and lowers hydrocarbon and toxic emissions.

51. E—Hub locations collect power and then distribute it to nearby locations that need it.

52. B—Fully 10% of our power comes from geothermal sources, mostly in the western United States.

Chapter 18

53. C—Sick building syndrome increases health problems and results in more sick days off.

54. D—To reduce fuel consumption and emissions, the CAFE standard requires vehicles to have an average fuel efficiency of 27.5 miles per gallon.

55. A—Total organic carbon is used by hydrologists to check the health of freshwater.

56. A—A solution with a pH value of 7 is neutral, while a pH value <7 is acidic and a pH value >7 is basic. Natural waters typically have a pH between 6 and 9.

Chapter 19

57. B—The greenhouse effect keeps the Earth warm enough for living things to survive.

58. C—Methane is a by-product of the production, transportation, and use of natural gas.

59. E—Ozone blocks ultraviolet radiation, which is harmful to living organisms.

60. D—Climate change takes place over a long time and is not the result of seasonal variations.

Scoring

On this Diagnostic Exam, the number of questions you answered correctly is your raw score.

Since the AP Environmental Science Exam also contains free-response questions, you'll need to practice those (see Sample Exam 2 at the back of the book) at a later time, but for now a rough approximation of your score can be obtained from the following.

RAW SCORE	ESTIMATED AP SCORE
35–60	5
26–34	4
19–25	3
11–18	2
0–10	1

Okay. How did it go? Did you score about where you thought you would? If you did, and your score was high, terrific! If you scored lower than you hoped, then congratulations! You've got the study information and time to improve your score. Environmental topics are increasingly important now.

Good luck!

Develop Strategies for Success

CHAPTER **4** Approaching Each Question Type

Approaching Each Question Type

IN THIS CHAPTER

Summary: A few reminders and tips on handling different question types to raise your AP score.

Key Ideas

Multiple-Choice Questions

- Read each question carefully.
- Try to answer the question before looking at the answer.
- Toss out wrong answers first.
- Draw a picture in the margin if it helps.
- Limit the time you spend on each question.
- Leave a question and move on if you have no idea. It won't count against you.

Free-Response Questions

- Print if your handwriting is tough to read.
- Be clear and try not to ramble. Organize your thoughts.
- Illustrate your answer with a graph or picture if it helps.
- Save time by giving one-word answers if possible.
- If a question asks *how* something happens, briefly explain *why* it is environmentally important.

Multiple-Choice Questions

As you probably know from years of test taking, there are special strategies used to answer each question type. Multiple-choice and free-response questions make the most of what you know.

Past experience also affects your confidence (positive or negative) on different test types. Personally, I like questions that have one answer. Period. Too many choices make me less, rather than more, comfortable.

I had an environmental science professor in college that wrote multiple-choice questions with the answers: (a) never true, (b) moderately true, (c) true, (d) often false, and (e) always false. Picking the right answer was excruciating and confused me so much I didn't feel fairly tested on my true knowledge level. However, some people are great at eliminating red herrings offered in multiple-choice questions and sail through without breaking a sweat. Here is how they do it. Multiple-choice questions have three parts:

1. The *stem* grounds the question. This can be presented as a fill-in-the-blank sentence, rather than a question.

   Example: It is projected that by 2015, a dozen cities in the world will have populations between __ and __ million residents. **Answer:** 15 and 30.

2. The *correct answer.* This is the single choice that best completes a sentence or answers a question.

   Example: What percent of the atmosphere is nitrogen? **Answer:** 79%.

3. *Distracters* (known in mystery stories as *red herrings*) take the reader in wrong directions. On tests, they are wrong answers thrown in to make students work for the right answer.

   Example: All the following are important water conservation methods except
   a. better farming techniques
   b. oscillating sprinkler systems
   c. dry cooling systems
   d. preventable runoff
   e. irrigation canals

   Answer: Oscillating sprinkler systems (all the answers are meant to seem reasonable).

We all learn and approach tests differently. Some students zero in on the right answer quickly. Others spot distracters, eliminate them, and then find the right answer like a sculptor chipping away at a block of granite. In this book I've included a few silly answer choices just for fun. Here are several tips to think about when taking the AP Environmental Science Exam.

1. *Read each question carefully.* Okay, it's a no-brainer, but it's easy to overlook a negative distracter or other test writing trick. Look at the following:

   Example: Which of the following is not a major metal used in the United States?
   a. Gold
   b. Nickel
   c. Sulfur
   d. Manganese
   e. Lead

   Answer: If a student zipped through this question, didn't see the *not,* and didn't know sulfur was a nonmetal, he or she might waste time rereading the question and trying to find the right answer. In this case, the word *which* points to a single choice and helps sort out the negative question. Just remember to watch for words like *not, most, least, always,* and *never.*

2. *Look before you leap* (or think before you look). If you know something about environmental science, you might know the right answer without looking. Terrific! Make use of that skill to save time and disappoint tricky test writers who have given a lot of thought to creating crafty distracters. Read the question, answer it in your mind, and *then* look at the possible choices before choosing the right one.

   Example: A ______________ is a species or group of species whose impact on its community or ecosystem is much larger and more influential than would be expected from simple abundance.

   Answer: Since you remember the answer is *keystone species,* you fill in the blank mentally before looking at the answer choices. This saves time, takes you quickly past distracters, and boosts your confidence.

3. *Relax. Some questions are just plain easy.* Believe it or not, some questions are straightforward and easy. If you get to the test and find a question that takes you 10 seconds to answer, don't jinx your answer with doubt because it's easy. Some answers are easy.

4. *To guess or not to guess.* A lot of students wonder whether or not they should guess. No points are deducted for wrong answers, so if you do not know the correct answer, go ahead and guess! You have a 1 in 5 (20%) chance of being right. If you can eliminate some answers, your odds of making the right choice go up. For example, if you can narrow the choices down to two, you have a 1 in 2 (50%) chance of picking the right answer.

5. *A picture is worth a thousand words.* Have you seen copies of Leonardo da Vinci's notes? He sketched out his ideas. Many students are very visual, and a quick drawing in the test book margin can make a question a lot easier to answer.

   Example: When a population first overshoots the carrying capacity of an environment, its numbers are

   a. at their highest
   b. at their lowest
   c. near the average
   d. not related to carrying capacity
   e. set for the remainder of the cycle

   When a species multiplies to the extent that it uses all local resources, it will experience dieback until resources are renewed and balance attained, as shown in Figure 4.1.

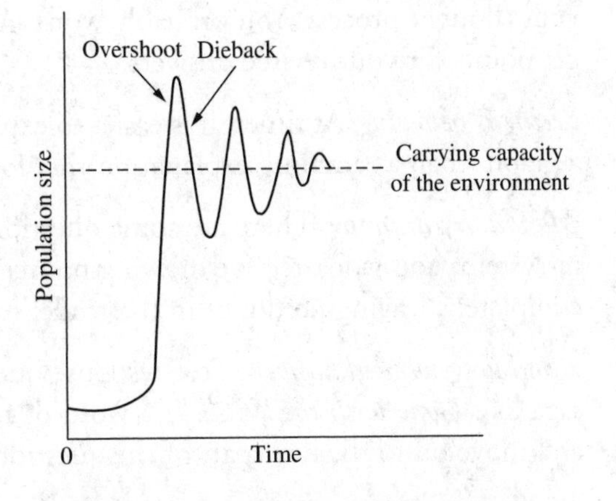

Figure 4.1 Population Overshoot

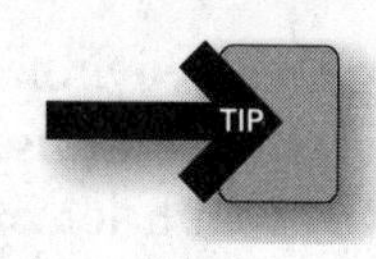

6. *Pace yourself.* Unlike the NASCAR races, there is not a pace car for AP exams. You have to do it yourself. If you don't know the answer to a question, skip it. You can always come back to it. It's better to move on. Remember, you only have a little over a minute (72 seconds) per question. You can still score well on an unfinished exam because unanswered questions are not counted against you. When you keep going, you raise your chances of finding and correctly answering more questions. Pay attention to your pace on the diagnostic and practice exams.
7. *Think, but don't dwell.* If you know the answer or can narrow down the choices, great. However, questions you spend too much time pondering are going to cost you time that could be used for "easy" questions. After working through *Five Steps to a 5: AP Environmental Science*, you will be prepared. Allow your mind to dropkick a tough question. If you have extra time, you can always go back to it.

8. *Worst case scenario.* Changing an answer is usually not a good idea. Overthinking and/or second guessing are often behind answer changing. Don't do it unless you have a *really* good reason to change. Plus, you take a chance of confusing the grading computer by poorly erasing your first answer.
9. *Fill them up!* Make sure you fill in answer circles fully. It only takes an extra couple of seconds and could make a difference. Also, every few questions, make sure you are filling in the answer circle to the matching question. It will be a disaster if you miscount. Definitely double check question numbers with answer circles when you turn a page.

Free-Response Questions

The score of the free-response questions is equal to one-half of your grade. Since it counts a lot, you have to know the material. Unlike multiple-choice questions, guessing from a list of choices isn't an option. However, there are tips to help you get every point possible from your answers.

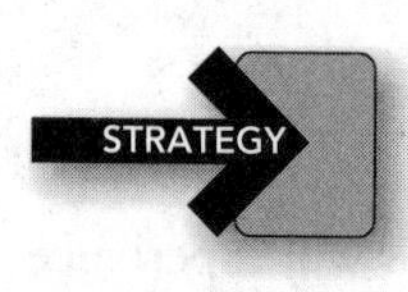

1. *Readability.* For a grader to give you all the points you deserve, he or she needs to be able to read your answers. So print if necessary, be organized, and make sure your answers make sense. If they do, you'll score higher.
2. *Follow through, even if you goof up.* Sometimes, when you're halfway through a question, you realize you made an error or bad assumption. If it's really huge, go back and fix it. If it's small, it's better to keep going (maybe with an explanation) so the reader knows your thought process. You probably won't get a point for the wrong answer, but you'll get points for your revised answers.
3. *Seeing is believing.* At times, it is easier to explain something with a sentence or two and a graph, than a rambling explanation that loses the reader.
4. *Mandatory drawing.* There are some questions that ask you to draw a structure, cycle, or system. You won't receive drawing points if you answer with text. Label your graph completely, leaving no doubt in the reader's mind that you understand the answer.

5. *Avoid long-winded answers.* Free-response questions are not your chance to write a long saga like *Gone with the Wind.* If a word or number answers a question, write it down and move on to the next part of the question.

6. *Capture gems.* Some questions ask how one variable affects another. For example, one population may decrease while another increases. If there is the possibility of receiving two points, don't just say "it decreases" (*how* it is affected), but add "due to the drain on mutually needed resources" (*why* it is affected).

7. *Pick and choose.* Free-response questions have several parts. If you don't understand or know the answer to one part, all is not lost. Keep going! Some parts can stand alone. If you goof up or just don't know something, find what you do know. You'll get points for all your right responses.

8. *Time in a bottle.* Budget your time. Spend a minute or two reading the question and planning your response.

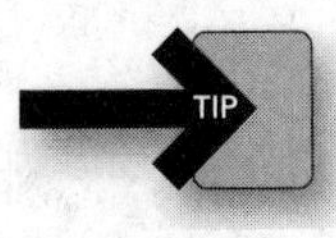

9. *Points, points, and more points.* Free-response questions have different parts worth 1 to 3 points each. If you answer all the parts correctly, you receive the maximum number of points. It is important to at least try to answer each part. You will not lose points and may gain one. However, if you skip a part and then write twice as much on the next part, you won't receive any more points than the maximum for that part.

10. *Circle the wagons.* Outlines are a good way to save time, prevent wandering answers, cover all the points, and write a well thought out response. Although you are not graded on the quality of your writing, anything that makes a reader's job easier makes him or her your friend.

11. *Memory game.* Sometimes your mind continues to work on a skipped question. Don't fight it. Keep going, answering what you can. If you have time at the end of the test and your memory has snagged the briefly blocked answer, go back and fill it in.

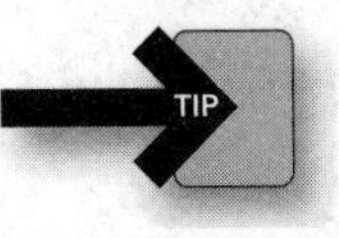

12. *Answer what is asked.* This seems obvious, I know, but some people try to overthink the question. Remember, with free-response questions you either know the answer (or some part of it) or you don't. Concentrate on answering what is asked. You'll save time and get more points overall.

STEP 4

Discover/Review to Score High

Earth Science Concepts

IN THIS CHAPTER

Summary: Land mass movements and environmental impacts of natural events have occurred over geological time.

Keywords

✪ Geological time, principle of uniformitarianism, plate tectonics, subduction, convection, earthquakes, P and S waves, volcanism

Investigating Our Environment

The Earth's limitless beauty and complexity provide broad areas for scientific study. Researchers from many different fields focus their skills on the mechanisms and interactions of hundreds of environmental factors. These natural and industrial factors affect the environment in ways that are known or suspected, as well as those that are totally unidentified. Some changes have been taking place for millions of years; others appear to be accelerating. Today, environmental scientists are sorting through tons of data to understand impacts of modern processes on the environment.

Geological Time

The concept of geological time describes time over millions of years. However, for scientists to be sure they are talking about the same time increment, they divide geological time into different amounts. These amounts are known as eons, eras, periods, and epochs. Table 5.1 shows the different time divisions.

Table 5.1 Geological time is divided into eons, eras, periods, and epochs.

EON	ERA	PERIOD	EPOCH	YEARS (MILLIONS)
Phanerozoic	Cenozoic	Quaternary	Holocene	0.1
			Pleistocene	2
		Tertiary	Pliocene	5
			Miocene	25
			Oligocene	37
			Eocene	58
			Paleocene	66
	Mesozoic	Cretaceous		140
		Jurassic		208
		Triassic		245
	Paleozoic	Permian		286
		Carboniferous		320
		Devonian		365
		Silurian		440
		Ordovician		500
		Cambrian		545

Time that spans millions of years is known as **geological time**. The entire history of the Earth is measured in geological time.

Geological time includes the history of the Earth from the first hints of its formation until today. It is measured mathematically, chemically, and through observation. Figure 5.1 shows a geological time clock with one second roughly equal to one million years.

In 1785, Scottish scientist James Hutton, often called the father of modern geology, tried to figure out the Earth's age. He studied and tested local rock layers in an attempt to calculate time with respect to erosion, weathering, and sedimentation.

Hutton knew that over a few years, only a light dusting of sediments are deposited in an undisturbed area. He figured out that sedimentary rock must have been compressed, tighter and tighter, from the weight of upper rock layers over many ages. He also thought that changes in a sedimentary rock layer, through uplifting and fracturing of weathering and erosion, could only have taken place over a very long period of time. Hutton was one of the first scientists to suggest that the Earth is extremely ancient compared to the few thousand years that earlier theories suggested. He thought the formation of different rock layers, the building of towering mountains, and the widening of the oceans must have taken place over millions of years.

Hutton wrote the *principle of uniformitarianism* suggesting that changes in the Earth's surface happened slowly instead of all at once. His early work paved the way for other geologists to realize the Earth was not in its final form, but still changing. Gradual shifting and compression changes occurred across different continental land forms.

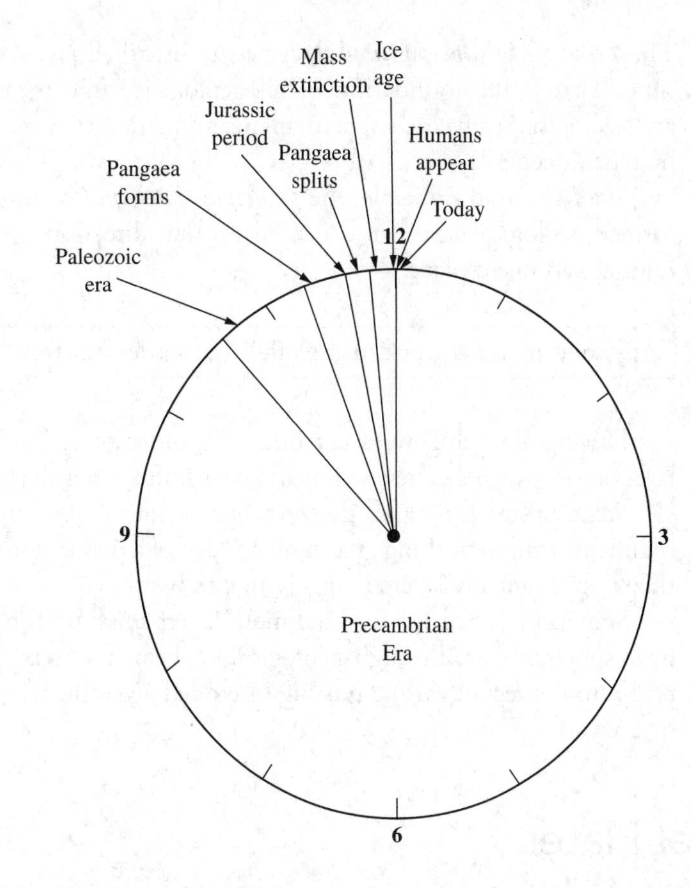

Figure 5.1 Compared to ancient eras, our modern time period is very small.

Time Measurements

Until the 17th century, people believed the Earth was approximately 6,000 years old. This estimate, based on humankind's history, was handed down through stories and written accounts. Except for theory, there were no "scientific" ways to check its accuracy. However, in the 1800s, following the early work of geological pioneers like Nicolaus Steno and Hutton, scientists began to test rock samples for their age. It was during this time that scientists began to use dating methods which suggested that the Earth was millions or even billions of years old.

Relative Time

Relationships between earth layers and soil samples are described by *relative time measurements,* which reveal age. With relative time measurements, the age of rock and soil layers is found by comparing them to neighboring layers above and below.

Dating an unknown sample to a certain time period when compared to samples of known ages is called *relative dating.*

In 1669, Nicolaus Steno came up with three ways to order samples in geological time:

1. Principle of superposition
2. Principle of original horizontality
3. Principle of lateral continuity

The *principle of superposition* describes undisturbed layers piled on top of each other over time. This is the foundation of all geological time measurements. For example, when archaeologists study ancient settlements and cities, they record the top layers first, followed by older, deeper layers.

Steno's second principle, the *principle of original horizontality*, described how sediments formed geological layers in a flat, horizontal direction even after base layer bending and folding had taken place.

Any solid material (rock or organic) that settles out from a liquid is known as *sediment.*

Driving along highways cut into hills and mountainsides, you will see horizontal rock layers at steep angles. These sediment layers shifted after the original sedimentation took place.

Steno's third principle, the *principle of lateral continuity*, describes how water-layered sediments thin to nothing at a shore or edge of a deposition area. This happens even though they were originally layered equally in all directions.

Sometimes scientists find sediment layers missing from different sections. These layers have split far apart through geological movements or by timeless erosion. If a sample is taken from a section with a missing or eroded layer, the true sedimentation picture wouldn't be seen.

Continental Plates

Geologists know that the original supercontinent, Pangaea, broke up into huge land masses (North America, Africa, etc.) called *continental plates.* There are 15 to 20 major plates of different sizes that make up the Earth's crust.

A geological *plate* is a layer of rock that drifts slowly over the supporting, upper mantle layer.

Continental and ocean plates range in size from 500,000 to about 97 million kilometers (km) in area. Plates can be as much as 200 miles (mi) thick under the continents and beneath ocean basins. Plates as much as 100 km thick fit loosely together in a mosaic of constantly pushing and shoving land forms. At active continental plate margins, land plates ram against other continental plates causing rock to push up into towering mountains.

The border between the Eurasian and Indian–Australian plates is a good example of plate conflict. Along this plate margin lies the Himalayan range with the world's tallest mountain (Mount Everest). Where the Nazca ocean plate and South American continental plate collide, the Andes Mountains formed. Similarly, where two ocean plates collide, one dives beneath the other and deep ocean trenches are formed.

Plate Tectonics

The outer crust, or *lithosphere,* is a puzzle of movable parts that mold to each other according to different applied pressures. Originally, scientists studied mountains, valleys, volcanoes, islands, earthquakes, and other geological happenings independently. Each study was considered unique and unrelated to other geological sites until widespread travel and communication began.

Geophysicist J. Tuzo Wilson put it all together. Knowing the movement and folding of the Earth's outer layers were ongoing, he devised the concept of *plate tectonics.* Continental drift made sense to Wilson when combined with the idea of large-plate movement and pressures.

Plate tectonics (*tektonikos* is Greek for "builder") describes the formation and movement of ocean and continental plates.

Plate tectonics is an umbrella theory, which explains the Earth's activity and the creation, movement, contact, and flattening of the lithosphere's solid rock plates.

Plate Movement

After Wilson's description of plate tectonics, geologists began comparing plate measurements and found that most plates aren't even close to their original positions! Fossils of tropical plants, once at the equator, were found in Antarctica. Deep rock in the Sahara desert, sliced by glacier travel, was frosty long before arriving at its current hot and dry retreat.

Most importantly, plates slide along at rates of up to 8 inches per year in some areas. The Pacific and Nazca plates are separating as fast as 16 centimeters (cm) per year, while the Australian continental plate is moving northward at a rate of nearly 11 cm per year. In the Atlantic, plates crawl along at only 1 to 2 cm per year.

When enough new material from within the Earth is deposited, plates slant, slide, collide, and push over, under, and alongside their neighbors. Continental and ocean plates ride over or dive under each other, forcing movement down into the mantle and liquid core.

A *subduction zone* is an area where two crustal plates collide and one plate is forced under the other into the mantle.

Figure 5.2 illustrates the subduction zone between plates. The lithospheric plate causes volcanism on the overriding plate, while the crustal plate is forced deep into the mantle.

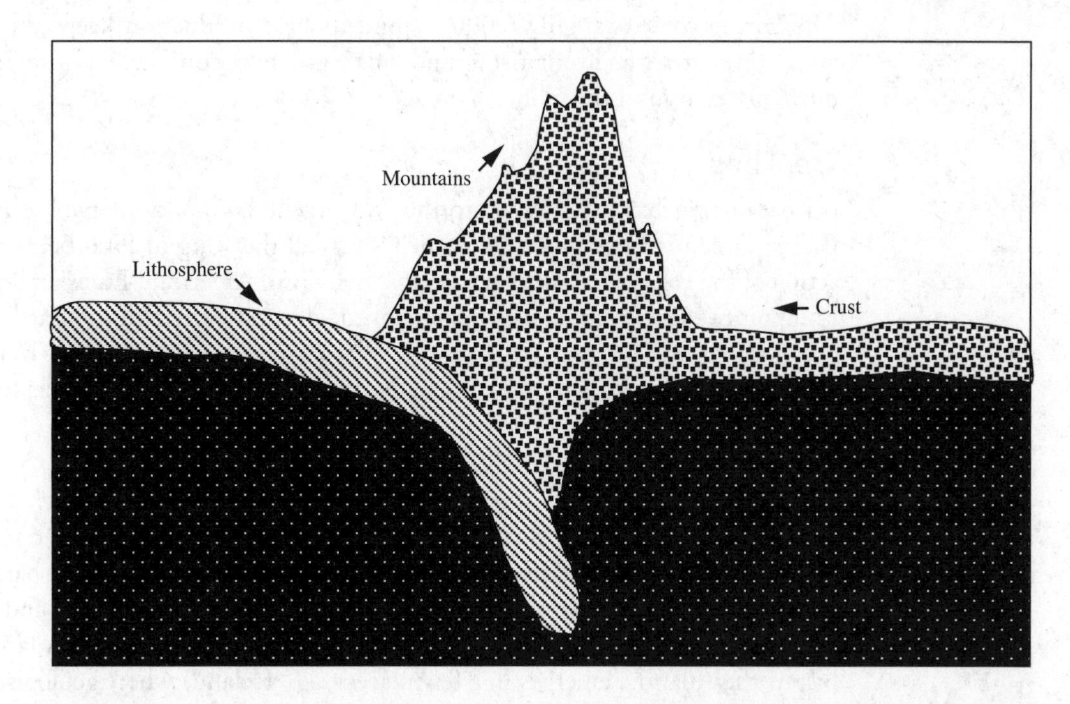

Figure 5.2 Mountain building occurs at subduction zones between plate margins.

Convection

Material circulation caused by heat is called *convection.* In the Earth's system, convection is affected by gravitational forces within the planet, as well as heat and radioactive recycling of elements in the molten core.

> *Convection* is the process of heat transfer that causes hot, less dense matter to rise and cool matter to sink.

All tectonic processes within the Earth involve movement of solid or malleable matter. Mantle convection, driven by the thermal gradient between the core and crust, takes place by the extremely slow deformation (creep) of the rocks and minerals in the upper and lower mantle and the transition zone. Flaws in mineral crystalline structures provide gaps. The application of extreme pressure causes atoms in the structure to creep, one atom at a time, into a new position.

Magma movement depends on temperature and density differences within large "pockets" of molten matter. Depending on conditions, magma rises in hotter pockets and sinks in cooler pockets. The Earth's molten core keeps thermal activity and tectonic processes going.

Plate Boundaries

TIP

There are three main types of plate boundaries that form when convection drives plates together and apart. These boundaries include

- *Convergent boundaries.* Plates clash, and one is forced below the other, pulling older lithosphere into the depths of the mantle. (See the Trenches section that follows.)
- *Divergent boundaries.* Plates pull apart and move in opposite directions making room for new lithosphere to form from outpouring magma. (See the Ridges section that follows.)
- *Transform fault boundaries.* Plates slide past each other parallel to their shared boundaries.

The Earth's core is roughly 6,000°C, and its heat production keeps the convection cycle going. Heat transfer from rising and sinking convection currents provides the power to move plates around the globe.

Trenches

An ocean trench is formed along the convergent boundary of two colliding plates. The Pacific Ocean is ringed by these trenches (called the Ring of Fire) because of the constant action of the Pacific oceanic plates against the North American, Eurasian, Indian–Australian, Philippine, and Antarctic plates. Some of the deepest points on the Earth are found within ocean trenches. The Java trench in the West Indies, and the Mariana Trench (five times as deep as the Grand Canyon) in the Pacific, average between 7,450 meters (m) and 11,200 m, respectively.

Ridges

Ridges are formed along divergent boundaries where plates slowly edge apart from each other. When plates separate, magma rises into the crack, filling it and hardening into rock. Most magma exiting the mantle today is found at ocean floor ridges and along plate margins. Nearly all ridges are at the bottom of oceans, but the Mid-Atlantic Ridge, which bisects the Atlantic, emerges in a few places (e.g., Iceland) where geologists can measure its growth and movement.

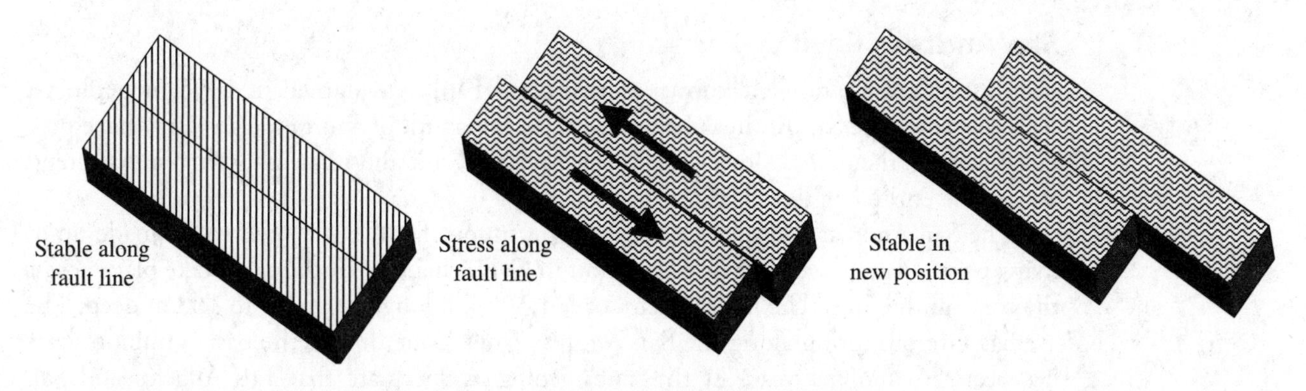

Figure 5.3 Stress builds along a fault until a plate suddenly snaps into a new position.

Transform Fault Boundaries

A *fault* is a crack between two tectonic plates caused by building pressures, which eventually causes surrounding rock to split.

Some plates slide past each other horizontally as in a *transform fault.* The rock on either side moves in opposite directions as the buildup of pressure between the plates provides energy. The well-known San Andreas Fault in California where the North American and Pacific plates meet is a transform fault boundary. Along this fault, the Pacific Ocean plate is sliding north while the continental plate is moving southward. Since these two plates have been at it for millions of years, the rock facing each other on either side of the fault is of different types and ages.

As with most plate collisions, transform faults do not slide along at a constant rate, but in fits and starts. Extreme friction is caused by the buildup of pressure between two grinding plates. This pressure is usually released by earthquakes and a sideways slip between the transform fault fractures. The 1906 and 1989 earthquakes near San Francisco were caused by side-slip transform fault movement.

Following a slip, pressure builds up for many years until it again reaches a critical point. Figure 5.3 shows stress buildup and fault displacement in a transform fault. One day, when the "last straw" is added by pressure buildup, everything shifts violently again. This sudden movement causes millions of dollars in damages to populated areas by breaking roads, building foundations, bridges, and gas lines.

Faults

Rocks, under enough pressure, reach a breaking point. As stress overwhelms rock binding forces something gives and an earthquake is triggered. A *fault plane* is used to represent an actual fault or a section of a fault. Faults are generally not perfectly flat, smooth planes, but fault planes give geologists a rough idea of a fault's direction and orientation. The intersection of a fault plane with the Earth's surface along a crack is called the *fault line* or *surface trace.* Faults lines are not always obvious on the surface.

A fault *trend* is the direction it takes across the Earth's surface. A *fault strike* is the line formed by the intersection of the fault plane with a horizontal plane. The direction of the strike is at an angle off due north. So, when discussing the direction of a certain strike, geologists might call it a northwest-striking fault.

San Andreas Fault

California's active and well-known San Andreas Fault was named in 1895 by geologist A.C. Lawson. The San Andreas Lake, about 32 km south of San Francisco, may have provided Lawson the name. However, the San Andreas Fault and its neighboring faults extend almost the entire length of California.

The San Andreas Fault is not a single, continuous fault, but a *fault system* made up of many parts. Plate movement is so common in California that earthquakes take place across the zone all the time. The fault system, over 1,300 km long, is also up to 16 km deep. The average rate of motion along the San Andreas Fault Zone during the past 3 million years has been 56 mm per year. At this rate, geologists speculate that Los Angeles and San Francisco will be side-by-side in about 15 million years.

Fault Size and Orientation

Active faults have a high potential for causing earthquakes. *Inactive faults* slipped at some point in time (causing earthquakes) but are now stuck solid. However, if local tectonic processes change, it is possible for inactive faults to reactivate.

Faults can measure from less than a meter to over a thousand kilometers in length with corresponding widths. Large fault depths are limited by the thickness of the Earth's crust.

$\uparrow$ fault slip area $\Rightarrow$ $\uparrow$ earthquake produced

It is important to remember that the size of a fault rupture is directly proportional to the size of the earthquake produced by the slip. When a set of fractures is large and developed, it is known as a *fault zone.*

Earthquakes

An *earthquake* describes the sudden slip of a fault between two plates. The associated ground shaking and vibrations are known as *seismic waves.* Volcanic activity and other geological processes may also cause stress changes that may result in an earthquake.

Seismic waves are caused by vibrations in the Earth from cracks or shifts in the underlying rock.

Earthquakes occur globally, but some regions are more earthquake prone than others. Earthquakes happen in all types of weather, climate zones, seasons, and at any time of day, making it impossible to predict exactly when one will occur. *Seismologists* examine historical earthquake activity and calculate the probability of an earthquake happening again.

Not all fault movement results in seismic waves strong enough for a person to feel. Minor shifts aren't even picked up by sensitive instruments. However, every shift results in some crustal movement along a fault surface.

Hypocenters and Epicenters

Since earthquakes are usually caused by tectonic activity, hypocenters are always located at some depth underground. Figure 5.4 illustrates hypocenters versus epicenters.

The *hypocenter* of an earthquake is the location beneath the Earth's surface where a fault rupture begins.

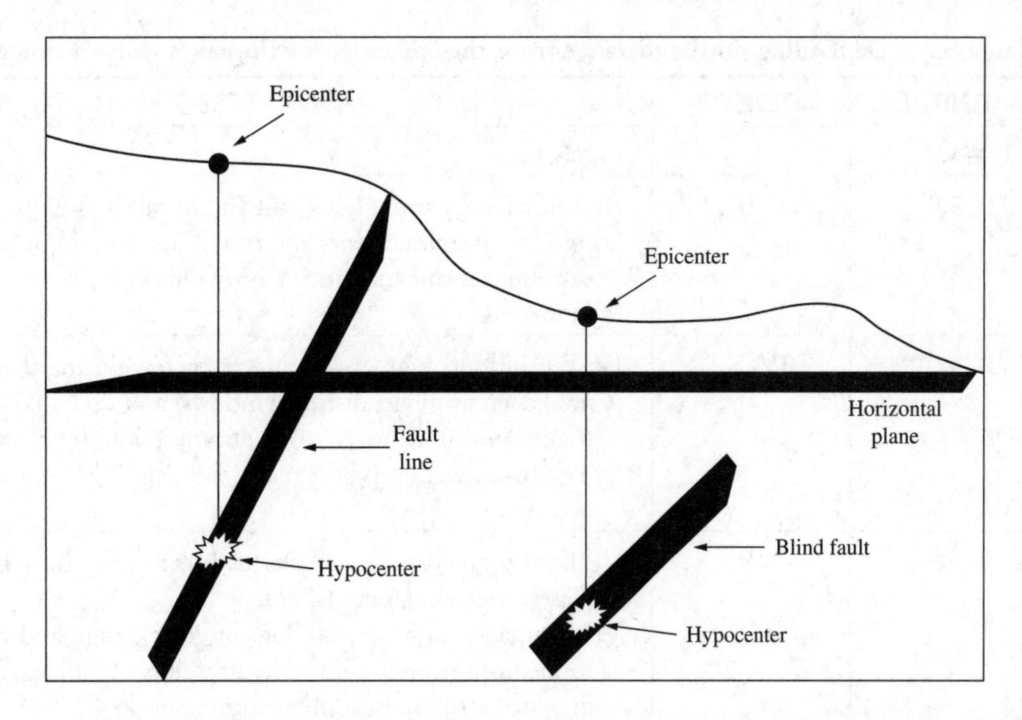

Figure 5.4 The epicenter is the surface point above the earthquake's hypocenter.

During TV earthquake coverage, the *epicenter* is often identified. This point is often named after local geography or a town closest to the epicenter (e.g., the San Francisco earthquake).

The *epicenter* of an earthquake is the location directly above the hypocenter on Earth's surface.

The hypocenter and epicenter may be far apart. The epicenter is found by comparing the distance from different measuring stations in a network. A deep hypocenter oriented at an angle away from vertical would not be found along the fault line, but at some distance away.

Earthquake Intensity

Seismology is based on the measurement and observation of ground motion. A pendulum *seismograph* was developed in 1751, but it wasn't until 1855 that geologists realized fault slips caused earthquakes.

A *seismograph* is an instrument that records seismic waves (vibrations) onto a tracing called a *seismogram.*

Richter Magnitude Scale

Seismographs record a zigzag trace of the changing amplitude of ground movement beneath the instrument. Sensitive seismographs greatly magnify these vibrations and can detect earthquakes all over the world recording time, location, and earthquake magnitude.

Table 5.2 Depending on the distance from the epicenter, earthquakes cause a range of damage.

MAGNITUDE	INTENSITY	EFFECTS
1.0–3.0	I	Rarely felt.
3.0–3.9	II–III	II: Only felt by those lying still and in tall buildings. III: Felt by people indoors and in tall buildings, but not recognized as an earthquake; cars may rock a bit; vibrations like truck going by; time estimated.
4.0–4.9	IV–V	IV: Felt indoors by many, outdoors by few during the day; some awakened at night; dishes, windows, and doors jarred; walls crack and pop; sounds like heavy truck ramming building; cars rocked a lot. V: Felt by almost everyone; many wake up; dishes and windows broken; objects tipped over.
5.0–5.9	VI–VII	VI: Felt by everyone, many alarmed; heavy furniture moved; some fallen plaster; overall damage slight. VII: Damage minor in buildings of good design and construction; slight to medium in well-built structures; lots of damage in poorly built or designed structures; some chimneys broken.
6.0–6.9	VIII–IX	VIII: Damage minor in specially designed structures; lots of damage and some collapse in common buildings; huge damage in poorly built structures; chimneys, factory stacks, columns, monuments, and walls fall; heavy furniture flipped. IX: Damage great in specially designed structures; well-designed frame structures jerked sideways; great damage in substantial buildings, with partial collapse; buildings moved off foundations.
7.0 and higher	X and higher	X: Some well-built wooden buildings destroyed; most masonry, frame and foundation destroyed; rails bent. XI: Few masonry structures left standing; bridges destroyed; rails bent completely. XII: Damage complete; lines of sight and level are distorted; objects airborne.

The *Richter Magnitude Scale* was developed in 1935 by Charles F. Richter of the California Institute of Technology to compare earthquakes. Using high-frequency data from nearby seismograph stations, Richter could measure an earthquake's magnitude (size). Each logarithmic magnitude rise in the Richter scale represents a tenfold increase in energy. Table 5.2 lists increasing earthquake magnitudes and effects.

Earthquakes with a magnitude of 2.0 or less are often called *microearthquakes.* Most people don't feel them and only nearby seismographs record their movement. Earthquakes of 4.5 or greater magnitudes (thousands each year) are recorded by sensitive seismographs worldwide. On average, one large earthquake of >8.0 occurs somewhere in the world yearly. In 2011, 19,000 lives were lost with $574 billion in structural damages, including those to a nuclear power plant, when Japan was hit by an offshore 9.0 magnitude earthquake.

Seismic Waves

Today's digital monitoring instruments use waveforms to record and analyze seismic data. In 2011, 19,000 lives were lost with $574 billion in structural damages, including those to a nuclear power plant, when Japan was hit by an offshore 9.0 magnitude earthquake.

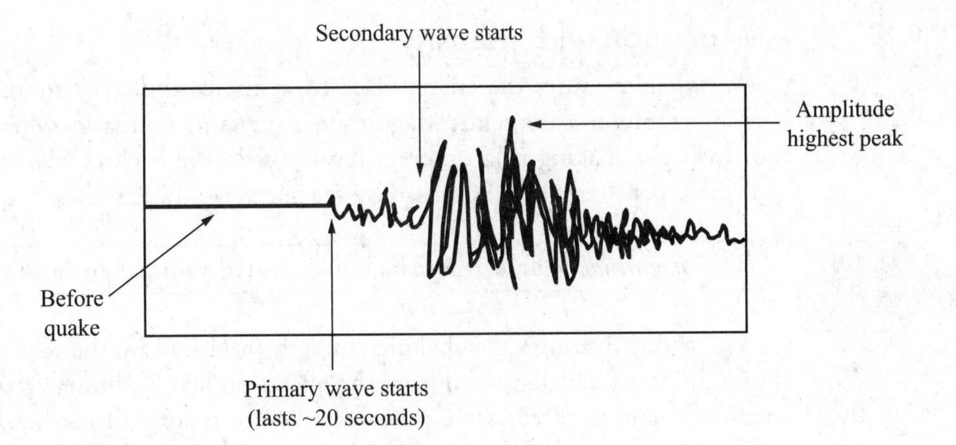

Figure 5.5 Seismographs record earthquake primary waves, secondary waves, and wave amplitude.

Resulting from a 1,200 km long section of the Indian Ocean floor sliding under the Southeast Asian Continental Plate, the seismic waves rattled the Earth for days. Over 300,000 people from 40 nations were reported dead or missing from the resulting tsunamis that hit Indonesia, India, and western Africa. Millions were homeless.

Body Waves

There are two types of seismic waves, *body waves* and *surface waves.* Body waves travel through the Earth's interior, while surface waves travel the top surface layers. Most earthquakes take place at depths of less than 80 km below the Earth's surface. Body waves are further divided into *P waves* and *S waves.* Figure 5.5 shows a typical earthquake tracing.

- *Primary waves.* Primary (P) waves, which bunch together and move apart like an inchworm, are the fastest seismic waves, moving 5 km/s (14 times faster than sound waves move through air) to arrive at a monitoring station. P waves are *longitudinal compression waves.* Like sound waves, they travel through rock or buildings, causing squeezing and expanding parallel to the direction of transmission. The speed of a P wave depends on the types of matter through which it moves. Generally, the denser the matter, the faster the P wave travels.
- *Secondary waves.* Secondary (S) waves travel around 60% of the speed of P waves and arrive at monitoring stations later. S waves are *transverse shear waves.* They cause a shearing, side-to-side motion perpendicular to their travel direction. Because of this, they can only travel through material (e.g., rock) with shear strength. Liquids and gases have no shear strength, so S waves can't travel through water, air, or molten rock.

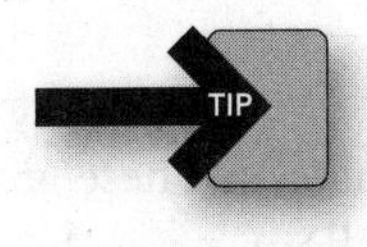

Since a seismogram indicator jumps wildly with P wave action, it can be tricky to spot lower frequency and longer wavelength S wave arrival. A sudden wavelength increase and amplitude jump indicates an S wave. However, when a large earthquake occurs, this isn't obvious, because P wave action hasn't slowed to the point where S waves overwhelm it.

Surface Waves

Surface waves are further divided into *Love waves* and *Rayleigh waves.* These waves travel along the Earth's surface and produce distinct types of motion. Love waves produce motion at a 90 degree angle to the wave's direction. This causes the horizontal shearing that wipes out building foundations.

Rayleigh waves produce a lazy rolling motion, causing buildings to experience a bobbing motion transverse to or parallel to a wave's direction of travel. This rolling, elliptical action is extremely hard for non-earthquake-proof buildings to withstand.

Magnitude and Intensity

Magnitude measures the energy released at the earthquake's source and is determined by seismograph measurements. Magnitude is the same no matter where you are or how strong or weak the shaking is in different locations. In the Richter Magnitude Scale, the amount of movement (amplitude) caused by seismic waves determines its magnitude.

The *magnitude* of an earthquake is a measured value of an earthquake's size.

Intensity measures the shaking strength produced by the earthquake at a certain location. Intensity is determined from the effects on people, human structures, and the natural environment. Richter's scale provides accurate reports of earthquake intensity. Table 5.2 gives intensities that are often seen near the epicenter of different earthquake magnitudes.

Types of Earthquakes

Oregon, Washington, and California sit on the North American Plate. The Juan de Fuca Plate is west of the Pacific Northwest coastline in the Pacific Ocean. The shared margin of these two plates is called the Cascadia Subduction Zone and is located 80 km offshore. When the Juan de Fuca Plate clashes with the North American Plate, it subducts into the Earth's mantle producing earthquakes.

Subduction Zone Earthquakes

When an oceanic plate is shoved beneath a continental plate, it may stick instead of sliding. This sticking causes pressure to build up until it is released suddenly as a major earthquake.

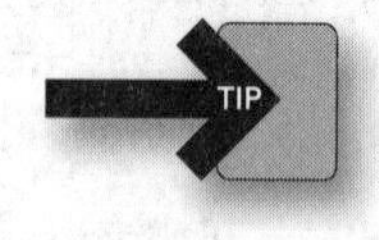

No large earthquakes have been recorded in the Cascadia Subduction Zone since records began in 1790, but the Cascadia had magnitude 8 to 9 earthquakes in the past. If an earthquake took place in this subduction zone, it would probably be centered off the coast of Washington or Oregon where the plates converge. The 1980 eruption of Mt. St. Helens may signal growing tectonic activity in the area.

Volcanoes

Magma movement is called *igneous* activity (from the Latin *ignis* meaning "fire"). Magma erupting from a volcano is called *lava,* which cools to volcanic rock.

A *volcano* is a mound, hill, or mountain formed from hot magma exiting the crust and piling up on the land or beneath the seas.

When referring to volcanic materials ejected during an eruption, most volcanologists talk about volcanic dust, ash, cinders, lapilli, scoria, pumice, bombs, and blocks. These are mostly sorted by size and composition and collectively called pyroclastic fallout.

Tephra describes all the different types of matter sent blasting from a volcano compared to slow flowing lava.

Mount St. Helens

Mount St. Helens was once considered one of the most beautiful mountains in the Pacific Northwest's Cascade mountain range. Then, in March of 1980, a series of increasingly stronger earthquakes (over 170) occurred before the mountain top was blown off in a tremendous explosion sending a column of volcanic ash and steam into the sky.

Volcanic ash ejected during a volcanic eruption is made up of rock particles less than 4 mm in diameter. Coarse ash is sized from $\frac{1}{4}$ to 4 mm, with fine ash (dust) measuring $<\frac{1}{4}$ mm in grain size.

Then, after two months of nearly constant tremors, a large bulge appeared on the mountain's side. On the morning of May 18, 1980, a huge earthquake a mile underground lowered the internal magma pressure within the volcano and caused the bulge to collapse, followed by a huge landslide moving nearly 250 km/h. The pressurized magma below exploded with a violence calculated at over 500 times the force of the Hiroshima bomb and sent ash 19 km into the air. Over 540 million tons of ash was spewed from Mount St. Helens over a 35,410 km^2 area, with most of it dropping on Oregon, Washington, and Idaho. As much as 3 inches of ash coated the countryside.

Hot Spots

Volcanoes, formed away from plate boundaries, are often a result of geological *hot spots.* A hot spot on the earth's surface is commonly found over a chamber of high-pressure and high-temperature magma. For example, the temperature of the basalt lava of the Kilauea volcano in Hawaii reaches 1,160°C.

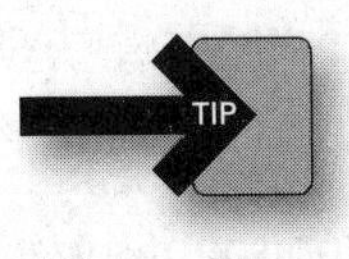

Volcanoes form when vents rise to the surface from a magma chamber. As a plate experiences magma upwelling from deep, ocean rifts, the volcano above slowly slides past the hot spot. The hot spot then begins forming a new volcano. Figure 5.6 shows how this process works. Repeated eruptions form a hot spot creating a string of closely spaced volcanoes. Island chains like the Hawaiian Islands were formed over a hot spot in the mantle.

Currently, a new island called *Loihi* is being formed at the end of the Hawaiian Island chain over the hot spot. Its peak is 1,000 meters below the water's surface and will break the surface in the next 10,000 to 100,000 years.

Craters

A bowl-shaped crater is found at the summit of most volcanoes and is centered over the vent. During an eruption, lava blasts or flows from the vent until the pressure below is released. The lava at the tail end of the eruption cools, sinks back onto the vent, and hardens. The crater fills with debris and sometimes its sides collapse. When this happens repeatedly,

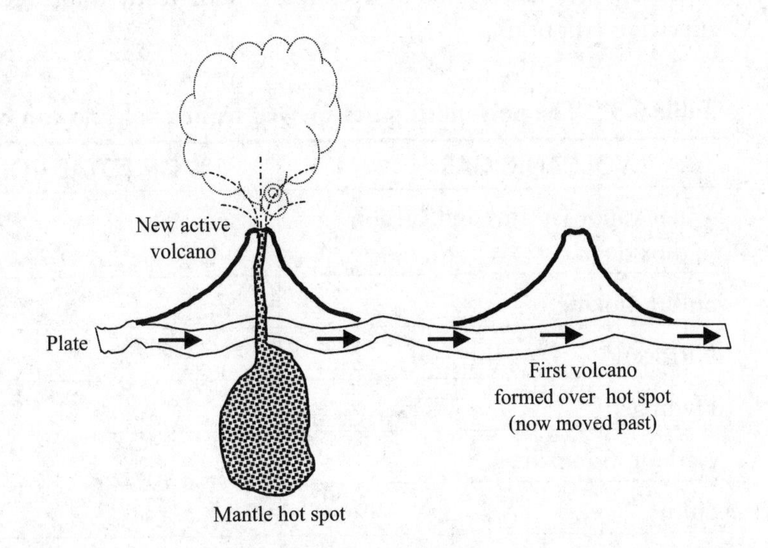

Figure 5.6 As the crust moves over a hot spot, a volcano is formed and eventually forms an island.

the crater grows to many times the size of the base vent and hundreds of meters deep. The crater on Mount Etna in Sicily is more than 300 m across and over 850 m deep.

Calderas

After a massive volcanic eruption spews huge volumes of lava from a magma chamber located a few kilometers below the vent, the empty chamber might not be able to support the weight of its roof. When this happens, the rock above the chamber caves in and a steep-walled basin, larger than the original crater is formed. Known as a *caldera,* these basins can be huge, ranging from 4 to 5 kms to more than 50 km across.

Crater Lake, Oregon, in the northwestern United States is formed from a caldera that is 8 km in diameter. Geologists think the original eruption, which caused the formation of the caldera, happened about 6,600 years ago.

Volcanic Gases

Since Earth's formation, elemental gases have played a part in its makeup and matter. Volcanologists collect eruption gases and sample lava flows at great risk to study their composition. Most eruption gases are made of water vapor from groundwater, seawater, or the atmosphere. Other gases such as sulfur dioxide, nitrogen, hydrogen, and carbon monoxide are from chemical changes in magma and rock during melting and release to the surface. Table 5.3 lists these gases.

TIP

Since most of these gases are poisonous in large quantities, people are often not killed by pyroclastic fallout and lava during an eruption, but by the searing hot, poisonous gases, that make breathing impossible. Archaeologists determined that when Mount Vesuvius erupted, most people were killed by the foul volcanic gases released.

Volcanic Activity

Volcanoes are classed into three types: active, dormant, or extinct. A volcano is *active* if it has erupted within recent recorded history. A *dormant* volcano has little erosion and may become active at any time. However, if a volcano has not erupted within recorded time and is greatly eroded, it is thought to be *extinct* and very unlikely to erupt again.

An average volcano erupts once every 220 years, but 20% of all volcanoes erupt only once every 1000 years and 2% erupt less than once every 10,000 years. There are nearly 600 active volcanoes on the planet today, with 50% of those erupting yearly. Luckily, volcanologists' use of modern technology has greatly improved the odds of predicting and surviving eruptions.

Table 5.3 The poisonous gases spewed from a volcano can be as dangerous as the lava.

VOLCANIC GAS	% OF TOTAL VOLCANIC GAS (AVERAGE)
Water vapor (steam) and carbon dioxide	90–95%
Sulfur dioxide	<1%
Nitrogen	<1%
Hydrogen	<1%
Carbon monoxide	<1%
Sulfur	<0.5%
Chlorine	<0.2%

❯ Review Questions

Multiple-Choice Questions

1. Archaeologists record the top layers first, followed by older, deeper layers to study ancient settlements and cities following the

(A) principle of original horizontality
(B) principle of uniformitarianism
(C) principle of superposition
(D) principle of uniformity
(E) principle of lateral continuity

2. Because he figured out that sedimentary rock must have been compacted and compressed, over many ages, _______ is known as the father of modern geology.

(A) Richard Palmer
(B) James Hutton
(C) W. Clayton Scott
(D) Nicolaus Steno
(E) Aubrey Hough

3. Relationships between earth layers and soil samples are described by

(A) tectonic time measurements
(B) Jurassic time measurements
(C) chronostatic time measurements
(D) continental drift
(E) subjective time measurements

4. When Mount Vesuvius erupted in 79 A.D., most people were killed by

(A) lava
(B) pyroclastic flow
(C) volcanic rock
(D) poisonous volcanic gases
(E) the hysterical crowd running toward the sea

5. The umbrella theory explaining the Earth's movement, contact, and flattening of large land plates is known as

(A) the Coriolis effect
(B) plate tectonics
(C) hot spots
(D) the Richter Magnitude Scale
(E) the subduction zone

6. Water-layered sediments that thin to nothing at a shore or the area where first deposited are the subject of the

(A) principle of uniformity
(B) principle of original horizontality
(C) second law of thermodynamics
(D) principle of superposition
(E) principle of lateral continuity

7. A volcano with little erosion and that may erupt at any time is classified as

(A) active
(B) dormant
(C) high risk
(D) sedimentary
(E) extinct

8. Vibrations in the Earth caused by cracks or shifts in the underlying rock are called

(A) caldera
(B) sedimentation
(C) igneous activity
(D) faults
(E) seismic waves

9. Where two crustal plates collide and one plate is forced under the other into the mantle, it is called a

(A) hot spot
(B) pressure zone
(C) surface trace
(D) subduction zone
(E) sedimentary layer

10. The principle of uniformitarianism includes the idea that

(A) continental shifting and compression are possible across different regions
(B) energy can neither be created nor destroyed
(C) volcanoes can erupt even when quiet for 200 years
(D) population growth is slow and the birth rate drops to around the death rate
(E) primary consumers (i.e., herbivores) consume plants and algae

11. A huge layer of rock that drifts slowly over the supporting upper mantle layer is a

(A) caldera
(B) hot spot
(C) geological plate
(D) tropical oasis
(E) transform fault

12. Any solid material (rock or organic) that settles out from a liquid is known as

(A) sediment
(B) ash
(C) a biofuel
(D) granite
(E) cinders

13. Love waves are a type of

(A) primary seismic wave
(B) surface wave
(C) pyroclastic flow
(D) folded rock
(E) lapilli

14. The following are all time units of the geological time scale except

(A) eons
(B) eras
(C) ages
(D) periods
(E) epochs

15. Gravitational forces within the Earth, as well as heat and radioactive element recycling in the molten core, take place through

(A) subduction
(B) deflection
(C) transformation
(D) convection
(E) divergent boundaries

16. An ocean trench is formed along the

(A) transform fault boundary of two colliding plates
(B) Mid-Atlantic Ridge
(C) distant edge boundary of two colliding plates
(D) divergent boundary of two colliding plates
(E) convergent boundary of two colliding plates

17. Primary waves, which bunch together and then move apart like an inchworm, are

(A) the slowest seismic waves
(B) the fastest seismic waves
(C) known as Rayleigh waves
(D) a type of surface wave
(E) the only wave to move through air

18. The location directly above an earthquake's hypocenter on the earth's surface is the

(A) epicenter
(B) trench
(C) hypercenter
(D) surface trace
(E) ridge

19. A large earthquake occurs, on average, somewhere in the world each year of greater than

(A) 1.5 magnitude
(B) 2.0 magnitude
(C) 4.0 magnitude
(D) 5.2 magnitude
(E) 8.0 magnitude

› Answers and Explanations

1. **C**—The principle states that younger strata are deposited on top of older layers during normal deposition.

2. **B**

3. **C**—Chronostatic time measurements are sequential.

4. **D**—During an eruption, searing hot and poisonous gases make breathing impossible.

5. **B**—Contact, grinding, and flattening of large land plates is known as tectonics.

6. **E**—Originally layered equally in all directions, the Grand Canyon's walls show this.

7. **B**—A dormant volcano hasn't erupted lately but may become active at any time.

8. **E**

9. **D**—Subduction zones can cause earthquakes and volcanism.

10. **A**—Hutton thought surface changes happened slowly and the land was still changing.

11. **C**

12. **A**

13. **B**—The motion is at a 90° angle to wave direction and causes horizontal shearing.

14. **C**—*Ages* is more a literary term for time than scientific.

15. **D**—Convection is affected by gravitational forces, as well as heat and radioactive recycling of elements in the molten core.

16. **E**—The clashing edges also cause volcano formation.

17. **B**—The speed of P waves depends on the types of matter they move through (i.e., the denser the matter, the faster a P wave moves).

18. **A**—An epicenter is found by comparing the distance from various measuring stations.

19. **E**

Free-Response Questions

1. On February 4, 1975, the Chinese government issued an immediate earthquake warning to the area of the city of Haicheng and began a huge evacuation. Nine hours later, Haicheng experienced an earthquake of magnitude 7.3. Luckily, most of the population was outside when 90% of the city's buildings were severely damaged or destroyed. Injuries were few.

Although there have been other prediction successes in this region of China, the misses are frequent. On July 28, 1976, a 7.8 magnitude earthquake hit Tangshan, 150 km east of Beijing, and home to over one million people. The hypocenter was located directly under the city, at a depth of 11 kilometers. About 93% of all buildings were destroyed and 240,000 people killed. Though the area was monitored, Chinese seismologists were caught off guard. In 2008, an 8.0 magnitude earthquake took place in China that was the second deadliest since Tangshan. It was centered in the Sichuan province and killed an estimated 70,000 people, leaving over 5 million people homeless.

(a) How are earthquakes caused?
(b) What is the difference between P waves and S waves?
(c) Should people be allowed to build in earthquake-prone areas?

2. Currently, a new island called Loihi is being formed at the end of the Hawaiian Island chain. Its peak is 1,000 meters below the water's surface and will break the surface in the next 10,000 to 100,000 years.

   (a) What is this process commonly called?
   (b) What causes this process?
   (c) Does this process affect the size of previously formed islands?
   (d) How does this process affect magma pressure deep in the Earth?

Free-Response Answers and Explanations

1.

   a. Continental plates are in constant motion. In areas where two plates come in contact the motion isn't smooth. As with most plate collisions, transform faults do not slide along at a constant rate, but in fits and starts. Extreme friction is caused by the buildup of pressure between the two grinding plates. This pressure is usually released by earthquakes and a sideways slip between the transform fault fractures.
   b. Though both types of body waves, P and S waves can be distinguished by their speed and seismic motion. P, or primary, waves, move quickly through the Earth's interior. They are longitudinal compression waves, moving like an inchworm parallel to the direction of transmission. S, or secondary, waves move more slowly and are transverse shear waves causing side-to-side motion within the Earth's interior.
   c. Policy is divided on this. In poor areas, people build where they own or lease land despite the potential danger. In many earthquake-prone cities, there are building code regulations in place to protect inhabitants during an earthquake.

2.

   a. Volcanoes, formed away from plate boundaries, are often a result of geological hot spots.
   b. Volcanoes form when vents rise to the surface from a magma chamber. As a plate experiences magma upwelling from deep, ocean rifts, the volcano above slowly slides past the hot spot. The hot spot then begins forming a new volcano.
   c. No. The other islands in the chain are finished forming.
   d. A hot spot on the Earth's surface is commonly found over a chamber of high-pressure and high-temperature magma. Magma pressure is released as the volcano flows up and out.

› Rapid Review

- Any solid material (rock or organic) that settles out from a liquid is known as sediment.
- Magma movement is called igneous activity (from the Latin *ignis* meaning "fire").
- The original supercontinent, Pangaea, broke up into huge land masses called continental plates.
- A fault is a crack between two rock plates caused by extreme, built-up pressure.
- An earthquake's hypocenter is the location beneath the Earth's surface where a fault rupture begins.
- The epicenter is the location on the surface that is directly above the hypocenter.
- The hypocenter and epicenter may be far apart.
- Tephra describes the matter blasted from a volcano compared to slow-flowing lava.

- When plates clash along convergent boundaries, one is forced below the other, pulling older lithosphere (crust) to deep mantle depths.
- A subduction zone is an area where two crustal plates collide and one plate is forced under the other into the mantle.
- An ocean trench is formed along the convergent boundary of two plates colliding with one another.
- Along divergent boundaries, plates pull apart and move in opposite directions making room for new lithosphere to form at the lip from outpouring magma.
- Ridges are formed along divergent boundaries where plates slowly edge apart.
- Plates slide past each other, parallel to their shared edges, along transform fault boundaries.
- P waves or longitudinal compression waves travel through rock or buildings, causing squeezing and expanding parallel to the direction of transmission.
- Love waves produce motion perpendicular to the direction of wave travel in a horizontal orientation *only* causing horizontal shearing, which wipes out building foundations.
- Rayleigh waves produce a lazy rolling motion, which causes buildings to experience a bobbing motion transverse to or parallel to the wave's direction of travel.
- Magnitude measures the size of an earthquake, while intensity measures the shaking strength created by the earthquake at a certain location.
- The intersection of a fault plane with the Earth's surface, along which a crack occurs, is called the fault line or surface trace.
- Each logarithmic magnitude rise in the Richter scale represents a tenfold increase in energy.

Atmospheric Composition

IN THIS CHAPTER

Summary: The atmosphere, made up of different distinct layers, is constantly changing. These changes and air circulations make up and affect the world's weather and climate.

Keywords

✪ Atmosphere, aurora, wind chill, jet stream, barometer, temperature inversion, Coriolis effect, tornado, hurricane, cyclone, El Niño, solar intensity, latitude

---

Composition

Many people think of weather when they hear the word *atmosphere.* People tend to talk about the weather and how the heat, cold, snow, wind, or rain will impact their weekend plans. Often they turn to the local weatherperson for the local or long-range forecast. A *meteorologist* is a person who studies the weather and its atmospheric patterns.

Weather describes the atmosphere's condition at a given time and place with respect to temperature, moisture, wind velocity, and barometric pressure.

Atmospheric gases blanketing the Earth exist in a mixture made up of about 79% nitrogen (by volume), 20% oxygen, 0.036% carbon dioxide, and trace amounts of other gases. Air is the common name for this gaseous mix.

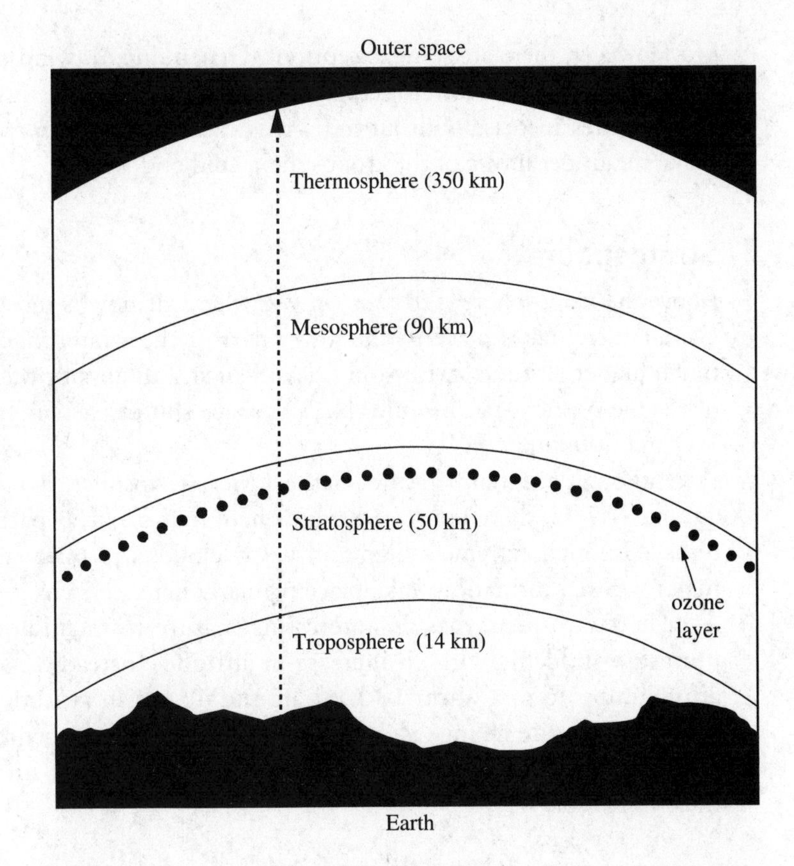

Figure 6.1 The atmosphere is divided into four main layers.

Atmospheric Layers

The atmosphere is divided into layers affected by gas mixing, chemical properties, and temperature. Nearest the Earth, the *troposphere* is about 8 km in altitude in the polar regions to 17 km around the equator. The layer above the troposphere is the *stratosphere,* reaching an altitude of around 50 km. The *mesosphere* stretches 80 to 90 km and lies above the stratosphere. Finally, the *thermosphere,* or *ionosphere,* is further out and fades to the black of outer space. Figure 6.1 illustrates these layers.

Troposphere

The atmospheric layer closest to the Earth's surface is the troposphere. Nearly all human activities occur in the troposphere, since living organisms are protected from the harmful cosmic radiation showers constantly raining down on the Earth's atmosphere.

The troposphere is where all the weather that we experience takes place. If you have ever survived a hurricane or tornado, you know it's an active place. Rising and falling temperatures, circulating air masses, and air pressure keep things lively.

When measured next to other layers, the troposphere is fairly slim, extending only 17 km up from the Earth's surface.

The *troposphere* is where all the local temperature, pressure, wind, and precipitation changes take place.

The warmest portion of the troposphere is found at the lowest altitudes. This is because the Earth's surface absorbs the sun's heat and radiates it back into the atmosphere. Commonly, as altitude increases, temperature decreases.

However, there are some exceptions. Depending on wind currents and the like, mountain ranges can cause lower areas in the troposphere to have just the opposite effect. When temperatures increase with altitude, it is called a *temperature inversion.* The wind speeds make the upper limits of the troposphere cold and windy.

Stratosphere

Above the troposphere is the *stratosphere* where air flow is mostly sideways. Most commercial air travel takes place in the lower part of the stratosphere. Military aircraft travel at much higher altitudes, with some classified, stealth aircraft thought to graze the boundary of the mesosphere and beyond. NASA's space shuttles generally travel to altitudes between 160 and 500 km.

Although the temperature in the lower stratosphere is cold and constant, hovering around −57°C, there are strong winds here that occur as part of specific circulation patterns. Although extremely high and wispy clouds can form in the lower stratosphere, no major weather formations take place regularly here.

The stratosphere has an interesting feature from midlevel on up. Its temperature jumps up suddenly with an increase in altitude. Instead of a frosty −57°C, the temperature jumps up to a warm 18°C at around 40 km in altitude in the upper stratosphere. This temperature change is due to increasing ozone concentrations that absorb ultraviolet radiation. The melding of the stratosphere upward into the mesosphere is called the *stratopause.*

Mesosphere

Above the stratosphere is the *mesosphere,* a middle layer separating the lower stratosphere from the inhospitable thermosphere. Extending from 80 to 90 km and with temperatures to around −101°C, the mesosphere is the intermediary of the Earth's atmospheric layers.

Thermosphere

The mesosphere changes to the *thermosphere* at a height of around 80 km. The thermosphere has rising temperatures that can reach an amazing 1,982°C. Thermospheric temperatures are affected by high or low sun spot and solar flare activity. The greater the sun's activity, the more heat is generated in the thermosphere.

Extreme thermospheric temperatures are a result of ultraviolet radiation absorption. This radiation enters the upper atmosphere, grabbing atoms from electrons and creating positively charged ions. This ionization gives the thermosphere its other name, the *ionosphere.* Because of ionization, the lowest area of the thermosphere absorbs radio waves, while other areas reflect radio waves. Since this area decreases and disappears at night, radio waves bounce off the thermosphere. This is why far distant radio waves can often be received at night.

Electrically charged atoms build up to form layers within the thermosphere. Before modern satellite use, this thermosphere deflection was important for long-distance radio communication. Today, radio frequencies able to pass through the ionosphere unchanged are selected for satellite communication.

The thermosphere is where the *aurora* resides. The *Aurora Borealis* and *Aurora Australis,* also known as the northern and southern lights, are seen in the thermosphere. When solar flares hit the magnetosphere (the region directly above the thermosphere) and pull electrons from atoms, they cause magnetic storms near the poles. Red and green lights are seen when scattered electrons reunite with atoms, returning them to their original state.

Seasons

The Earth's rotation around the sun, combined with its axial tilt, allows for different seasons: summer, winter, fall, and spring. Depending on the region, different amounts of solar energy strike the Earth at various times of the year. The equator gets the most, and the poles the least. Distance from the equator and the intensity of solar radiation have a direct effect on seasonal changes.

Latitude

Warm equatorial air vapor moves northward, cools, and condenses to fall as rain or snow depending on the season. Heat and moisture are distributed in global circulation patterns from the equator to the northern latitudes. Vertical convection currents, known as *Hadley cells,* have low-pressure circulation and rising air. High-pressure circulation cells occur where air sinks.

Solar Intensity

A *solar radiation unit* of measurement is 1 langley, which is equal to 1 calorie per square centimeter of the Earth's surface (i.e., 3.69 British thermal units [Btu] per square foot). Solar radiation is also stored in materials like water and soil.

Incoming solar energy also evaporates water into vapor. As the water changes form, it releases stored energy known as *latent heat.* Then, when water vapor reverts to liquid form, it releases 580 calories of heat energy. When this takes place over large bodies of water like an ocean, evaporation converts huge amounts of solar energy into latent energy.

Coriolis Effect

In physics and math, the *Coriolis effect* is a perceived deflection of moving objects viewed from a turning frame of reference. The Coriolis effect was named in 1835 after a French mathematician, Gustave Gaspard Coriolis, who published a set of equations explaining how objects acted in theoretical rotating systems. Although his research was not applied to the atmosphere, it explained directional winds across the globe. The nearly constant easterly winds that dominate most of the tropics and subtropics are known as trade winds. This is an optical illusion such that air moving from the north pole seems to turn right (northern hemisphere) and left (southern hemisphere) due to the Earth's rotation. This is *not* a result of the Earth's curvature or gravitation, but of rotation. Figure 6.2 illustrates the Coriolis effect and wind rotation.

Jet Stream

When watching the evening weather report, chances are good you will hear something about the *jet stream.* This speedy current is commonly thousands of kilometers long, a few hundred kilometers wide, and only a few kilometers thick. Jet streams are found between 10 to 14 km above the Earth's surface in the troposphere. Blowing from west to east at speeds of 240 km/h, they can also dip northward or southward depending on atmospheric conditions.

The *jet stream* is a long, narrow current of fast moving air found in the upper atmospheric levels.

Air temperature differences drive the jet stream. The bigger the temperature differences, the stronger the pressure differences between warm and cold air. Strong pressure differences create strong winds. This is why jet streams fluctuate so much in speed.

During the winter months, polar and equatorial air masses form a sharp surface temperature contrast causing an intense jet stream. The strong jet stream pushes farther south

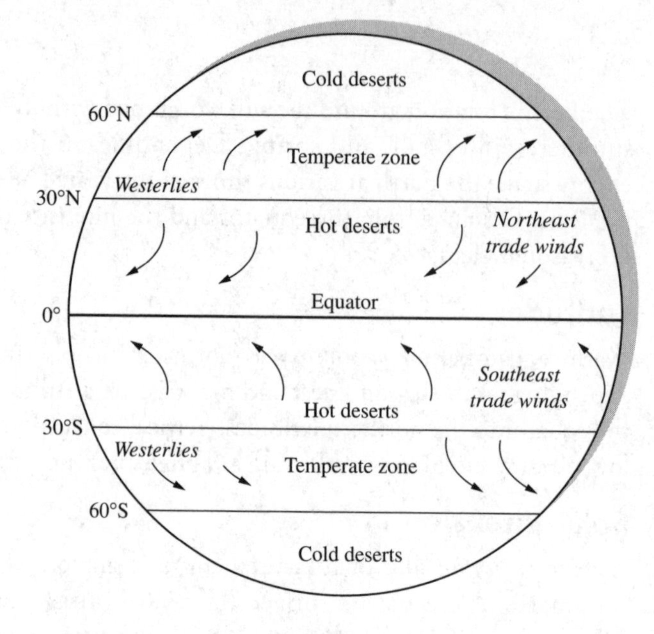

Figure 6.2 Trade winds are found at certain latitudes or regions on the Earth.

in the winter. However, during the summer months, when the surface temperature difference is less severe, jet stream winds are weaker. The jet stream then moves farther north.

Wind Chill

Wind chill factor measures the rate of heat loss from exposed skin to that of surrounding air temperatures.

Wind chill occurs when winter winds cool objects down to the temperature of the surrounding area; the stronger the wind, the faster the rate of cooling. For example, the human body is usually around 36°C in temperature, a lot higher than a cool Montana day in November. Our body's heat loss is controlled by a thin insulating layer of warm air held in place above the skin's surface by friction. If there is no wind, the layer is undisturbed and we feel comfortable. However, if a wind gust sweeps this insulating layer of air away, we feel

Table 6.1 Wind chill can bring down the temperature of the body quickly.

		TEMPERATURE (°C)							
		–15°C	–10°C	–5°C	0°C	5°C	10°C	15°C	20°C
Wind speed (km/h)	0	–15	–10	–5	0	5	10	15	20
	5	–18	–13	–7	–2	3	9	14	19
	10	–20	–14	–8	–3	2	8	13	19
	30	–24	–18	–12	–6	1	7	12	18
	50	–29	–21	–14	–7	0	6	12	18
	70	–35	–24	–15	–8	–1	6	12	18
	90	–41	–30	–19	–9	–2	5	12	18

chilled. The protective air layer must be reheated by the body. See Table 6.1 to get an idea of the wind chill equivalent temperatures at different wind speeds.

Air Pressure

Bakers living in the mountains have to consider air pressure when creating light cakes and soufflés. Lower pressure at high altitudes (over 6,000 km) changes the baking process from that of sea-level baking. In fact, cake mixes give different directions for high-altitude baking to make up for the pressure difference on the rising cake.

Air pressure is the force applied on you by the weight of air molecules.

Although air is invisible, it still has weight and takes up space. Free-floating air molecules are pressurized when crowded into a small volume. The downward force of gravity gives the atmosphere a pressure or a force per unit area. The Earth's atmosphere presses down on every surface with a force of 1 kilogram (kg) per square centimeter. The force on 1,000 cm^2 is nearly a ton!

Weather scientists measure air pressure with a *barometer.* Barometers measure air pressure in centimeters of mercury or *millibars.* A measurement of 760 mm of mercury is equal to 1013.25 millibars.

Air pressure tells us a lot about the weather. With a high-pressure system, there are cooler temperatures and sunny skies. When a low-pressure system moves in, look for warmer temperatures and thunderstorms.

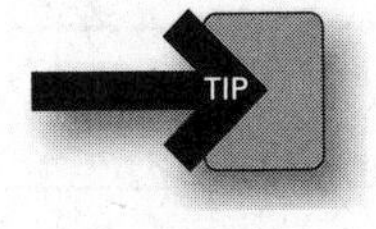

Atmospheric pressure falls with increasing altitude. A pillar of air in cross section, measured from sea level to the top of the atmosphere, weighs approximately 14.7 pounds per square inch (psi). Atmospheric pressure (atm) at sea level is equal to

1 atm = 760 mmHg (millimeters of mercury) = 1013 millibars
= 14.7 psi = 1013.25 hPa (hectopascals)

On weather maps, changes in atmospheric pressure are shown by lines called *isobars.* An isobar is a line connecting areas of the same atmospheric pressure. It's very similar to the lines connecting equal elevations on a topographical map.

Relative Humidity

Growing up in a dry western state where a humid day had 10% humidity and then moving to the Texas Gulf Coast with months of 100% *humidity* was a shock. But what is humidity, anyway? Humidity is the amount of water vapor in the air.

Relative humidity is the link between the air's temperature and the amount of water vapor it contains.

At any specific temperature, there is a maximum amount of moisture that air can hold. For example, when the humidity level is forecast at 75%, it means that the air contains $\frac{3}{4}$ of the amount of water it can hold at that temperature. When the air is completely saturated and can't hold any more water (i.e., 100% humidity), it rains.

Since the air's ability to hold water is dependent on temperature, hotter air holds more moisture. This temperature-dependent, moisture-holding capacity contributes to the formation of all kinds of clouds and weather patterns.

Tornadoes

As speeding cold fronts smash into warm humid air, a convection of temperature and wind is formed. Winds in tornadoes can easily reach speeds of over 250 km/h. Large tornadoes contain the fastest winds ever measured on the Earth and have been recorded at over 480 km/h.

Tornadoes are usually classified into one of the following three different levels:

1. *Weak tornadoes* (F0/F1) make up roughly 75% of all tornadoes. They cause around 5% of all tornado deaths and last approximately 1 to 10 minutes with wind speeds <180 km/h.
2. *Strong tornadoes* (F2/F3) make up most of the remaining 25% of all tornadoes. They cause nearly 30% of all tornado deaths and last 20 minutes or longer with wind speeds between 180 and 330 km/h.
3. *Violent tornadoes* (F4/F5) are rare and account for less than 2% of all tornadoes, but cause nearly 65% of all tornado deaths in the United States. They have been known to last for one to several hours with extreme wind speeds of 330 to 500 km/h.

In the late 1960s, University of Chicago atmospheric scientist T. Theodore Fujita realized that tornado damage patterns could be predicted according to certain wind speeds. He described his observations in a table called the *Fujita Wind Damage Scale.* Table 6.2 shows the Fujita scale used today with its corresponding wind speeds and surface damage.

Table 6.2 A tornado's strength is rated by the Fujita Wind Damage Scale.

TORNADO RATING	TYPE	SPEED	DAMAGE
F0	Gale	64–116 km/h (40–72 mph)	Light damage: some damage to chimneys, tree branches break, shallow-rooted trees tip over and sign boards damaged.
F1	Moderate	117–180 km/h (73–112 mph)	Moderate damage: beginning of hurricane wind speeds, peels roofs, mobile homes moved off foundations or overturned, and moving cars shoved off roads.
F2	Significant	181–251 km/h (113–157 mph)	Considerable damage: roofs peeled, mobile homes smashed, boxcars pushed over, large trees snapped or uprooted, and heavy cars lifted off ground and thrown.
F3	Severe	252–330 km/h (158–206 mph)	Severe damage: roofs and walls torn off well-made houses, trains overturned, most trees in forest uprooted, and heavy cars lifted off ground and thrown.
F4	Devastating	331–416 km/h (207–260 mph)	Devastating damage: well-made houses leveled, structures blown off weak foundations, and cars and other large objects thrown around.
F5	Incredible	417–509 km/h (261–318 mph)	Incredible damage: strong frame houses are lifted off foundations and carried a considerable distance and disintegrated, car-sized missiles fly through the air in excess of 100 m, and trees debarked.
F6	Inconceivable	510–606 km/h (319–379 mph)	The maximum wind speed of tornadoes is not expected to reach the F6 wind speeds.

Tornadoes are unpredictable. Weather forecasters can tell when tornado conditions are ripe, but they don't know if or where they will strike.

Hurricanes

A *hurricane* starts as a series of thunderstorms over tropical ocean waters. To start, ocean water must be warmer than 26.5°C. The heat and water vapor from this warm water serves as the hurricane's basic fuel source.

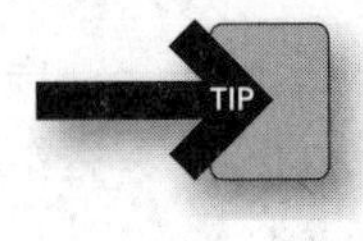

The first phase in the formation of a hurricane is the lowering of barometric pressure. This is called a *tropical depression.* In the next phase, the storm intensifies to a *tropical storm.* Favorable atmospheric and oceanic conditions affect the speed of the hurricane's development to the next step.

High humidity in the lower and middle troposphere is also needed for hurricane development. This high humidity slows cloud evaporation and increases heat released through increased rainfall. The concentration of heat is critical to driving the system.

Vertical wind shear affects a hurricane's development. During weak wind shear, a hurricane grows taller and releases condensed heat directly above the storm causing it to build.

Wind shear describes the sudden change in the wind's direction or speed with increasing altitude.

When wind shear is intense, heat is released and distributed over a larger area. Atmospheric pressure and wind speed change across the diameter of a hurricane. Barometric pressure falls quickly as wind speed increases.

The *eye* of the hurricane is the central point around which the rest of the storm rotates, and where the lowest barometric pressures are found.

The main feature most people look for on a weather map is the *eye* of the hurricane. The eye, roughly 20 to 50 km across, is found in the hurricane's center.

Just outside the hurricane's eye is the *eye wall* where the most intense winds and heaviest rainfall are found. Although wind speed in the eye wall is at its highest, at the eye, where barometric pressure is the lowest, winds are very light or calm. Remember, the winds are spinning constantly. Table 6.3 lists the different hurricane force categories.

Hurricanes have winds over 64 knots and turn counterclockwise about their centers in the Northern Hemisphere, and clockwise in the Southern Hemisphere. Course depends upon location. A hurricane in the eastern Atlantic is driven westward by easterly trade winds. These storms turn northwestward around a subtropical high and move into higher latitudes. As a result, the Gulf of Mexico and the eastern coast of the United States are at risk for hurricanes yearly.

Over time, hurricanes move into the middle latitudes and are driven northeast by the westerlies, merging with midlatitude fronts. Since hurricanes get their energy from the warm tropical waters, they fizzle quickly after moving over cold water or continental land masses.

Severe storms are known by different names around the world. Storms forming over the Atlantic or eastern Pacific Oceans are called hurricanes. In the northwestern Pacific Ocean and Philippines, these are *typhoons,* while Indian and South Pacific Ocean storms are known as *cyclones.*

Table 6.3 Hurricanes are rated according to specific strength categories.

HURRICANE CATEGORY	STRENGTH	WINDS AND STORM SURGE
1	weak	65–82 knot winds 1.2–1.7 m above normal storm surge
2	moderate	83–95 knot winds 1.8–2.6 m above normal
3	strong	96–113 knot winds 2.7–3.8 m above normal
4	very strong	114–135 knot winds 3.9–5.5 m above normal
5	near total devastation	>135 knot winds >5.5 m above normal

Since 1953, the Tropical Prediction Center has created lists of hurricanes names. As a tropical depression turns into a tropical storm, it's given the next name on the list. Written in alphabetical order, the names alternate between male and female. If a specific hurricane has been particularly vicious, the name is never used again. For example, Hurricanes Alicia, Andrew, Betsy, Camille, Carmen, Gilbert, Hugo, Katrina, Ike, and Roxanne have been retired from use.

El Niño–Southern Oscillation

First described by Sir Gilbert Thomas Walker in 1923, the *El Niño–Southern Oscillation* (*ENSO;* or simply El Niño) is defined as sustained sea surface temperature changes of greater than 0.5°C for longer than 5 months across the central tropical Pacific Ocean. Commonly, ENSO takes place every two to seven years and lasts approximately one to two years.

El Niño and La Niña come from the Spanish words for "little boy" and "little girl." El Niño is said to refer to the "Christ child" since the event is often seen off the west coast of South America in late December around Christmas time.

Normally, the Pacific pattern consists of equatorial winds bringing warm water westward, while cold water upswells along the South American coast. However, during an El Niño year, warm water nears the South American coast (without the cold upswell) and gets even warmer. La Niña, a mild version of El Niño, pushes warm water further west than usual.

The first sign of an El Niño is a rise in air pressure over the Indian Ocean, Indonesia, and Australia. Next, air pressure falls over Tahiti and the eastern or central Pacific Ocean. Trade winds in the south Pacific weaken or move east, followed by warm air rising near Peru, bringing rain to the northern Peruvian deserts. Finally, warm water spreads from the west Pacific and Indian Ocean to the eastern Pacific, causing widespread drought in the western Pacific and rain in the dry eastern Pacific.

Predicting this cyclical weather pattern in the Pacific, Atlantic, and Indian Oceans is important, since ENSO events have global impacts. For example, since yearly circulation patterns of cold, nutrient-rich ocean waters change, ENSO is connected with worldwide fishing problems. Temperature-dependent fish species increase or decrease according to biological requirements. Fish populations (e.g. Peruvian sardines and shrimp, normally sustained on the nutrient-rich cold waters) move southward to colder Chilean waters.

El Niño has been observed for over 300 years, but major events have taken place in 1790–93, 1828, 1876–78, 1891, 1925–1926, 1982–83, 1997–98, and 2009–10. The El Niño of 1997–98 was especially strong and warmed the air by –16°C compared to the usual –17.5°C increase. Scientists are monitoring whether these events are increasing in intensity or frequency during global warming.

❯ Review Questions

Multiple-Choice Questions

1. Which of the following is the highest layer of the atmosphere?

(A) Stratosphere
(B) Thermosphere
(C) Ozone
(D) Troposphere
(E) Mesosphere

2. Which two gases make up the majority of the Earth's gases?

(A) Oxygen and methane
(B) Oxygen and propane
(C) Nitrogen and carbon dioxide
(D) Benzene and nitrogen
(E) Nitrogen and oxygen

3. Large temperature differences and strong pressure differences between warm and cold air cause the formation of the

(A) precipitation
(B) auroras
(C) jet stream
(D) clouds
(E) fossilization

4. The relationship between air temperature and the amount of water vapor it contains is known as

(A) relative humidity
(B) indistinct humidity
(C) point source humidity
(D) aridity
(E) wind shear

5. A sudden change in the wind's direction or speed with increasing altitude is called

(A) wind stop
(B) wind chill
(C) troposphere
(D) wind shear
(E) ionosphere

6. A long, narrow, upper atmosphere current of fast moving air is known as a

(A) jet stream
(B) thermophile
(C) contrail
(D) gust of wind
(E) typhoon

7. A tropical depression is the first phase in the formation of a

(A) tornado
(B) cloud
(C) hurricane
(D) jet stream
(E) temperature inversion

8. In which layer do all the pressure, wind, and precipitation changes occur?

(A) Outer space
(B) Stratosphere
(C) Troposphere
(D) Ionosphere
(E) Mesosphere

9. Tornadoes of the F4/F5 category account for what percent of deaths?

(A) 80%
(B) 65%
(C) 40%
(D) 25%
(E) 10%

10. Most commercial aircraft travel takes place in the lower part of the

(A) upper troposphere
(B) United States
(C) inverted air mass above a mountain range
(D) stratosphere
(E) ionosphere

11. The northern and southern lights are found in the

(A) mesosphere
(B) tropopause
(C) stratosphere
(D) troposphere
(E) thermosphere

12. When moving through the eye wall of a hurricane and into the eye, surface pressure

(A) increases
(B) decreases
(C) stays the same
(D) is at its highest
(E) is only rarely low

13. The thermosphere is also called the

(A) mesosphere
(B) troposphere
(C) ionosphere
(D) lower latitude layer
(E) El Niño effect

14. ENSO takes place at intervals of 2 to 7 years and lasts approximately

(A) 6 months
(B) 1 to 2 years
(C) 3 years
(D) 2 to 5 years
(E) 7 years

15. A hurricane's course depends upon location and

(A) trade wind direction
(B) eye size
(C) speed
(D) name
(E) eye wall

16. The Coriolis effect is a result of the Earth's

(A) gravitation
(B) temperature
(C) rotation
(D) plant life
(E) curvature

17. An isobar is a

(A) measure of temperature
(B) line connecting areas of increasing atmospheric pressure
(C) measure of depth
(D) line connecting areas of the same atmospheric pressure
(E) topographical symbol

18. Using Table 6.1, if the wind is blowing 30 km/h and the outside temperature is 15°C, what is the wind chill factor in degrees Celsius?

(A) 8°C
(B) 10°C
(C) 12°C
(D) 14°C
(E) 15°C

19. Sea surface temperature changes greater than 0.5°C for greater than 5 months across the central tropical Pacific Ocean are known as

(A) an arctic effect
(B) the jet stream
(C) a typhoon
(D) El Niño–Southern Oscillation
(E) ocean bloom

❯ Answers and Explanations

1. B—The thermosphere has temperatures up to 1,982°C and is affected by sun spot and solar flare activity.

2. E

3. C—When warm and cold air clash, there is often violent weather.

4. A—The link between air temperature and water vapor (e.g., hotter air holds more water).

5. D—Vertical wind shear affects a hurricane's development causing it to build.

6. A—Jet streams are 10 to 14 km above Earth's surface in the troposphere.

7. C—High humidity in the troposphere and low barometric pressure also add to hurricane development.

8. C—Rising and falling temperatures, rotating air masses, and pressure keep things lively.

9. B

10. D—Military aircraft reach much higher altitudes with some grazing the mesosphere.

11. E—These are the Aurora Borealis (northern) and Aurora Australis (southern) lights.

12. B—Pressure is lowest at the center and is used to track a hurricane's movement.

13. C—From the many ions present from cosmic radiation.

14. B

15. A—Temperature and pressure are also important factors.

16. C—This is *not* a result of the Earth's curvature or gravitation but of rotation.

17. D—Weather maps show atmospheric pressure changes with isobars.

18. C—The body has to reheat its warm, protective air layer when winds blow it away.

19. D—El Niño–Southern Oscillation is often shortened to ENSO.

Free-Response Questions

1. The Peru Current, which supports vast populations of food fish, slows somewhat during normal years due to warming, but in El Niño years the warming lasts for months and has a much greater economic impact upon international fishing. An El Niño event started in September 2006 and lasted until early 2007. Then, from June 2007 on, a weak La Niña occurred, eventually strengthening in early 2008. The Chilean government placed restrictions on fishing to help local fisheries compete against international companies moving southward after fish.

(a) What is the first sign of an El Niño event?
(b) Why do El Niño events change precipitation patterns on the western coast of South America?
(c) What role does air pressure play in weather patterns?
(d) Why is it important to study El Niño trends and fish impacts?

Free-Response Answers and Explanations

1.

a. The first sign of an El Niño is a rise in air pressure over the Indian Ocean, Indonesia, and Australia.

b. Because El Niño events cause changes or reverses in air pressure in the eastern and western Pacific, the trade winds that normally carry warm waters away from South America's Pacific coast toward Australia fail to do so. The warmer waters evaporate more easily causing larger amounts of clouds, and thus rainfall.

c. Air pressure can tell us a lot about the weather. With a high-pressure system, there are cooler temperatures and sunny skies. When a low-pressure system moves in, look for warmer temperatures and thunderstorms.

d. ENSO is connected with worldwide fishing problems. Temperature-dependent fish species increase or decrease according to biological requirements. Fish populations (e.g., Peruvian sardines, shrimp) normally sustained on the nutrient-rich cold waters move southward to colder Chilean waters.

› Rapid Review

- Relative humidity is the connection between air temperature and the amount of water vapor the air contains.
- Air temperature differences cause the jet stream. The greater the temperature differences, the stronger the pressure differences between warm and cold air. Strong pressure differences create strong winds and the jet stream.
- Violent tornadoes (F4/F5) account for less than 2% of all tornadoes, but cause nearly 65% of all tornado deaths.
- The eye of the hurricane is the central point around which the rest of the storm rotates, and where the lowest barometric pressures are found.
- When temperatures actually increase with altitude, it is called a temperature inversion.
- The troposphere is where all the temperature, pressure, wind, and precipitation changes we experience take place.
- The first phase in the formation of a hurricane is the lowering of barometric pressure, called a tropical depression.
- The Coriolis effect is a result of rotation, not gravitation.
- The name *hurricane* is only given to systems that form over the Atlantic or the eastern Pacific Oceans. In the northwest Pacific Ocean and the Philippines they are called typhoons, while Indian and South Pacific Ocean storms are called cyclones.
- A meteorologist studies the weather and its atmospheric patterns.
- Although air is invisible, it still has weight and takes up space.
- Nitrogen and oxygen make up the majority of Earth's gases, even in the higher altitudes.
- ENSO takes place every 2 to 7 years, but can occur yearly.
- The worst winds and heaviest rainfall are found in a hurricane's eye wall, not the eye.
- A solar radiation unit equals 1 langley or 1 calorie/cm^2 of Earth's surface or 3.69 Btu/ft^2 (British thermal units per square foot).
- Trade winds are caused by solar heating and convection currents.

Global Water Resources and Use

IN THIS CHAPTER

Summary: Water is critical for life. Although much of the world's land has available water, as many as 1 billion people do not have access to safe water. Conservation and distribution of water resources is important to the future of agriculture and the well-being of the world's human and animal inhabitants.

Keywords

✪ Hydrology, reservoir, conservation, residence time, groundwater mining, runoff, infiltration, salinity, soil porosity, watershed, aquifer, water table

Global Water

Although humans live mostly on land, the Earth is a water planet with nearly 72% of its surface covered by 1.3 billion km^3 of water. Deep oceans give us food, trade items, trade routes, recreation, and entertainment. Our bodies are made up of 66% water and require more water daily. We developed in amniotic fluid and can't last over 2 to 3 days without water. We wash ourselves and nearly everything else in water, get much of our food from water (oceans), and travel on water.

Hydrology is the study of the occurrence, distribution, and movement of water on, in, and above the Earth.

Since the beginning of time, the Earth's water has been used over and over. The water you drink today may have once been part of a tropical cove with a *Brachiosaurus* family on the beach nearby. Thanks to water circulation that same water has probably been liquid, solid, and gas many times over geological time.

The Water Cycle

The oceans hold over 97% of the Earth's water, the land masses hold 3%, and the atmosphere holds less than 0.001%. Freshwater is stored in ice caps, glaciers, groundwater, lakes, rivers, and soil. Yearly global precipitation is estimated at more than 30 times the atmosphere's total ability to hold water. Water is quickly recycled between the Earth's surface and the atmosphere. Table 7.1 shows how long water stays in different locations before being recycled.

The oceans are a source for the atmosphere's evaporated moisture. Around 90% is returned to the oceans through rainfall. The remaining 10% is blown across land masses where temperature and pressure changes cause rain or snow. Water, not lost through evaporation and rainfall, balances the cycle through runoff and groundwater flowing back to the seas.

Atmospheric water is thought to be replaced every 8 days. Water in oceans, lakes, glaciers, and groundwater is recycled slowly, often over hundreds to thousands of years.

A water *reservoir*, in the atmosphere, ocean, or underground, is a place where water is stored for some period of time.

Some water resources (like groundwater) are being consumed by humans at rates faster than can be resupplied. When this happens, the water source is said to be *nonrenewable*.

Table 7.1 Water is stored for different time periods in different places.

WATER SOURCE	VOLUME (*thousands of km^3*)	TIME IN LOCATION (*average*)
Atmosphere	13	8–10 days
Rivers and streams	1.7	16 days
Swamps and marshes	3.6	Months to years
Soil moisture	65	14 days–1 year
Biological moisture in plants and animals	65	1 week
Snow and glaciers	29,000	1–10,000 years
Freshwater lakes	125	1–100 years
Saline lakes	104	10–10,000 years
Shallow groundwater	4,000	Days to thousands of years
Oceans	1,370,000	3,000–30,000 years
Aquifers and groundwater	4,000	10,000 years

Salinity

Salinity levels are written in parts per thousand (ppt) with 1,000 grams salt/1 kg of water. It's estimated that if all the oceans' water were drained, salt would cover the continents 1.5 meters deep. Ocean salinity is about 35 ppt and varies (32 to 37 ppt) as rainfall, evaporation, river runoff, and ice formation affects it. For example, the Black Sea is so diluted by river runoff its average salinity is around 16 ppt.

When salt water moves into the polar regions, it cools and/or freezes, getting saltier and denser. Since cold, salty water sinks, ocean salinity increases with depth.

Freshwater salinity is usually less than 0.5 ppt. Water between 0.5 and 17 ppt is called *brackish.* The salt content is too high to be drinkable and too low for seawater. In areas such as estuaries, where fresh river water joins salty ocean water, the water is brackish.

Ocean Circulation

Wind pushing the ocean's surface contributes to ocean currents. Differences in water density, salinity, and temperature also augment ocean circulation. Huge currents, called *gyres,* carry water north and south, redistributing heat between the lower and higher latitudes. For example, current flowing south from Alaska keeps San Francisco, California, and the northern Pacific coast cool and misty much of the year.

Redistribution of ocean water is also seen in the Gulf Stream, which carries warm Caribbean water north past Canada's eastern coast to northern Europe. This current contains around 800 times the volume of the Amazon, the world's largest river. When the warm Gulf Stream water gets to Iceland, it cools, evaporates, becomes denser and saltier, and then sinks, creating a deep, powerful, southbound current, which completes the cycle back to the Gulf.

Groundwater

Water found below the Earth's surface is known as *groundwater.* This water fills subterranean spaces, cracks, and mineral pores. Depending on the geology and topography, groundwater is either stored or flows toward streams.

It is sometimes easy to think of groundwater as all the water that has been under the land's surface since the Earth was formed; but remember that nearly all water is circulating through the hydrologic cycle. We will take a closer look at this in Chapter 10.

The *residence time,* or length of time that water spends in the groundwater portion of the hydrologic cycle, differs a lot. Water may spend as little as days or weeks underground, or as much as 10,000 years. Residence times of tens, hundreds, or even thousands of years are not unknown. Conversely, average river water turnover time, or the time it takes a river's water to completely replace itself, is roughly 2 weeks.

Groundwater is found in one of two soil layers. The layer closest to the soil's surface is called the *zone of aeration,* where spaces between soil particles are filled with both air and water.

Under this surface layer lies the *zone of saturation,* where all the open spaces have become filled with water. The depth of these two layers is often dependent on an area's topography and soil makeup.

The *water table* is found at the upper edge of the zone of saturation and the bottom edge of the zone of aeration.

The water table is the boundary between these two layers. As the amount of groundwater increases or decreases, the water table rises or falls accordingly. Figure 7.1 shows the location of the water table line in relation to other ground layers.

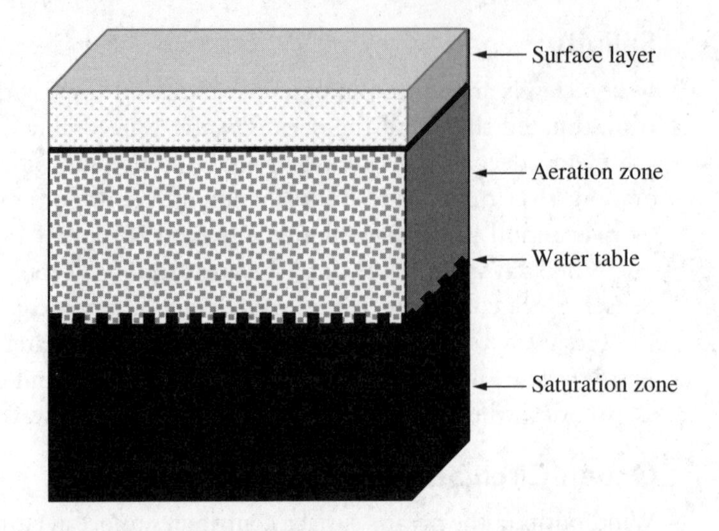

Figure 7.1 The water table is found below the surface between the aerated and saturated zones.

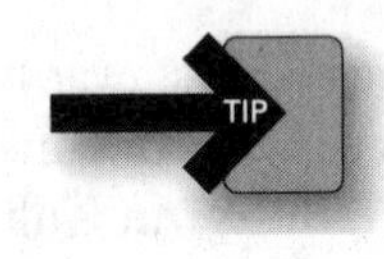

The amount of open space in the soil is called *soil porosity.* A highly porous soil has many gaps for water storage. The rate at which water moves through the soil is affected by *soil permeability.* Different soils hold different amounts of water and absorb water at different rates.

Hydrologists monitor soil permeability to predict flooding. As soil pores fill, more and more water is absorbed and there are fewer places for the extra water to go. When this happens, rainwater can't be absorbed by otherwise thirsty soil and plants and adds to flood waters. Flooding often happens in the winter and early spring, since water can't penetrate frozen ground. Rainfall and snow melt have nowhere to go and become runoff.

Watersheds

Streams and rivers get their water from hills, valleys, and mountains. The water runs downhill through streams or river beds. During hot months, these beds dry out, but are refilled by rain or melting snow later in the season.

The geographical region from which a stream gets water is called a *drainage basin* or *watershed.*

The boundary between two watersheds is called a *divide.* The Continental Divide in North America is the high line running through the Rocky Mountains. Rainfall and streams on the east side of the Rocky Mountains drain to the Atlantic Ocean or Gulf of Mexico, while flowing water from the western slopes of the Rocky Mountains runs to the Pacific Ocean.

Drainage basins are complex structures with thousands of streams and rivers draining from them, depending on their geographical size. Streams are described by the number of tributaries (other streams) that drain into them. Most streams follow a branching drainage pattern, which is known as *dendritic drainage.*

The United States Geological Survey (USGS) is the main federal agency that keeps records of natural resources and has counted approximately 1.5 to 2 million streams in the United States. These streams have increasingly larger areas of drainage and flow rates. A river is a wide natural stream of fresh water. The Mississippi River, a 10th-order stream, drains 320 million km^2 of land area.

A hydrologist determines flow rates for rivers and streams. This information, important in the design and evaluation of natural and constructed channels, bridge openings, and dams, is a critical factor in understanding water availability and drainage issues.

Aquifers

Groundwater is stored in *aquifers.* Aquifers are large underground water reservoirs. There are two main types of aquifers, *porous media aquifers* and *fractured aquifers.*

Porous media aquifers are made up of individual particles such as sand or gravel. Groundwater is stored in and moves through the spaces between the individual grains.

Permeable soil like sandstone includes lots of interconnected cracks or spaces that are large enough to let water move freely. In some permeable materials, groundwater moves several meters a day, while in other regions water flows just a few centimeters in 100 years.

Fractured aquifers, as the name implies, are made up of broken rock layers. In fractured aquifers, groundwater moves through cracks, joints, or fissures in otherwise solid bedrock. Fractured aquifers are often made of granite and basalt.

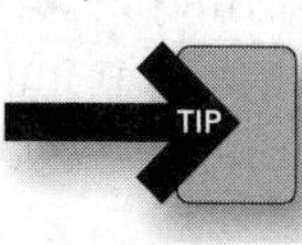

Rainfall soaks into the soil and moves downward until it hits impenetrable rock, at which time it turns and begins to flow sideways. Aquifers are often found in places where water has been redirected by an obstacle such as bedrock.

Groundwater in soil or rock aquifers within the saturated zone adds up to large quantities of water. Aquifers that form a water table separating unsaturated and saturated zones are called *unconfined aquifers* since they flow right into a saturated zone.

Groundwater doesn't flow well through impermeable matter such as clay and shale. Some aquifers, however, lie beneath layers of impermeable materials like clay. These are known as *confined aquifers.* These aquifers do not have a water table separating the unsaturated and saturated zones.

Confined aquifers are more complex than unconfined aquifers that flow freely. Water in a confined aquifer is often under pressure. This causes well water levels to rise above the aquifer's water level. The water in these reservoirs rises higher than the top of the aquifer because of the confining pressure. When the water level is higher than the ground level, the water flows freely to the surface forming a flowing artesian well.

A *perched water table* occurs when water is blocked by a low-permeability material below the aquifer. This disconnects the small, perched aquifer from a larger aquifer below. They are separated by an unsaturated zone and a second, lower water table.

Groundwater returns to the surface through aquifers that empty into rivers, lakes, and the oceans. Groundwater flow, with speeds usually measured in centimeters per day, meters per year, or even centimeters per year, is much slower than runoff.

When water flowing from an aquifer gets back to the surface, it is known as *discharge.*

Groundwater flows into streams, rivers, ponds, lakes, and oceans, or it may be discharged in the form of springs, geysers, or flowing artesian wells.

Water Consumption

Although many people act as if there is an unlimited amount of water available, supply in urban areas can get overloaded from increasing population needs. Withdrawal exceeds local supply, and existing water is often polluted or heated from industrial processes.

Consumption is the amount of withdrawn water lost in transport, evaporation, absorption, chemical change, or otherwise made unavailable as a result of human use.

As the global population has increased, human water demand has nearly doubled, especially in developing countries. Average worldwide water use is nearly 650 m^3 (175,000 gallons) per person per year. However, location is everything. In countries like Canada and Brazil with relatively low populations, the withdrawal rate is less than 1% of their annual renewable supply. In the United States, about 40% of the total annual renewable supply is withdrawn, while countries with relatively few freshwater resources like Israel withdraw 100% of their available supply each year.

Groundwater Mining

One issue concerning hydrologists is the depletion of water from aquifers faster than they can be refilled naturally. Humans are the main culprits in aquifer depletion, called *groundwater mining*. Huge volumes of groundwater are pumped out of aquifers for drinking water or irrigation. In fact, over 85% of withdrawn water is used for human and animal consumption in some areas, while 60% of withdrawn water is used for crop irrigation in others. In India, for example, over 90% of withdrawn water is for agricultural use.

When an aquifer is severely depleted in regions with few water resources, water must be rationed. Nonessential water use like lawn watering must be eliminated, and only water for consumption and hygiene allowed so that everyone has enough water to drink.

Aquifer Recharge

When water enters an aquifer, it is said to *recharge*. Groundwater often discharges from aquifers to replenish rivers, lakes, or wetlands. An aquifer may receive recharge from these sources, an overlying aquifer, or from rainfall or snow melt followed by infiltration.

Infiltration takes place when precipitation soaks into the ground and is strongly affected by soil porosity. Once in the ground, infiltrated water becomes groundwater and is stored in aquifers. However, if it is quickly pulled up to the surface, the levels can't build up.

An *aquifer recharge zone,* either at the surface or below ground, supplies water to an aquifer and/or most of the watershed or drainage basin.

Groundwater discharge adds considerably to surface water flow. During dry periods, streams get nearly all their water from groundwater. Physical and chemical characteristics of underground formations have a big effect on the volume of surface runoff.

While the rate of aquifer discharge controls the volume of water moving from the saturated zone into streams, the rate of recharge controls the volume of water flowing across the Earth's surface. During a rainstorm, the amount of water running into streams and rivers depends on how much rain an underground area can absorb. When there is more water on the surface than can be absorbed by the groundwater zone, it runs off into streams and lakes.

Runoff

Permeability is the measure of how easily something flows through a material. The higher the soil permeability, the more rain seeps into the ground. However, when rain falls faster than it can soak into the ground, it runs off.

Runoff is made up of rainfall or snow melt that has not had time to evaporate, transpire, or move into groundwater reserves.

Water always takes the path of least resistance, flowing downhill from higher to lower elevations, eventually reaching a river or its tributaries. All the land from which water drains to a common lake or river is considered to be part of the same watershed or runoff zone. Watershed drainages are defined by topographic divisions, which separate surface flow between two separate water systems.

Water runoff flows over land into local streams and lakes; the steeper the land and less porous the soil, the greater the runoff. Rivers join together and eventually form one major river, which carries the subbasins' runoff to the ocean.

Overland runoff also occurs in urban areas. Land use activities in a watershed affect the quality of surface water as contaminants are carried away by runoff and groundwater, especially through infiltration of pollutants. Because of this, hydrologists are concerned about surface pollutant runoff making its way into underground aquifers. Understanding factors affecting the rate and direction of surface and groundwater flow helps hydrologists determine where good water supplies exist and how contaminants migrate.

Agricultural, Industrial, and Domestic Use

Land use and soil treatment are linked with human activity. Pesticides and fertilizers used on crops affect water purity when runoff joins with surface and/or groundwater. Industrial water use is linked to the processing of hazardous chemicals. These may also impact streams or underground aquifers. Unfortunately, most water quality problems come from populated areas and improper land use.

Building and construction sites also contribute to runoff. Site preparation for new buildings and roads produce loose sediment. If not contained, this soil is washed away in a heavy rainfall along with pesticides, sewage, and industrial waste.

Additionally, septic systems, dumps, and landfills also bring pollutants into the overall water cycle. The United States has spent over $300 billion in the last 30 years on pollution control. However, there are still a lot of heavily polluted streams, rivers, and lakes.

Water is critical to life. Too much or too little water can have tremendous consequences. In the next 1,000 years, conservation and an acute understanding of the way planetary water is transported and stored will be essential.

Conservation

Water management and conservation are important in making water available for everyone. Environmentally frugal technology like low-water toilets, low-volume shower heads, reduced-use campaigns, and flood regulation can lower withdrawal and consumption. Planting native and drought-resistant landscapes and lawns lowers watering needs. A waterless toilet has been developed in Sweden that produces compost from human and kitchen waste. In 1998, cities such as Los Angeles, California; Austin, Texas; and Orlando, Florida, required all new building construction to have water-saving toilets to conserve water resources.

Large amounts of water can be reclaimed from treated wastewater and sewage. In California, over 30% of reclaimed water comes from treated sources. Purified and reclaimed water is being used for everything from crop irrigation to toilet flushing. Currently, California uses nearly 600 million m^3 of recycled water each year. Los Angeles uses roughly two-thirds of that amount for its annual water consumption.

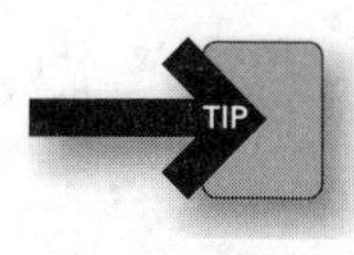

Globally, better farming techniques, irrigation canals, and preventable runoff and evaporation may reduce water requirements by 70%. Industrial water use such as engine and turbine cooling can be lowered or eliminated by installing dry cooling systems. As with other conservation and environmental concerns, cost is a factor. Some countries can't afford to change existing technology or pass costs on to local users as is done in industrialized countries.

Water is becoming limited for recreational use and wildlife as well. This is yet another aspect of a complex and important issue.

Science Community

Many scientists are concerned that changes in ocean ecosystems brought about by overexploitation, physical alteration, pollution, introduction of alien species, and global climate change are outpacing study efforts.

The scientific community's concern for our oceans compelled more than 1,600 marine scientists to sign the *Troubled Waters* statement in 1998. This statement, a project of the *Marine Conservation Biology Institute* in Washington, D.C., explains the problems involving oceans to policy makers and the public. It describes massive damage to deep-sea coral reefs from bottom trawling on continental plateaus and slopes, seamounts, and mid-ocean ridges. Complex ecosystems with thousands of marine species are destroyed by bottom trawling. Continued bottom trawling offers slim to no chance of recovery.

The *Troubled Waters* statement explains:

> To reverse this trend and avert even more widespread harm to marine species and ecosystems, we urge citizens and governments worldwide to take the following five steps:
>
> 1. Identify and provide effective protection to all populations of marine species that are significantly depleted or declining, take all measures necessary to allow their recovery, minimize bycatch, end all subsidies that encourage overfishing and ensure that use of marine species is sustainable in perpetuity.
> 2. Increase the number and effectiveness of marine protected areas so that 20% of Exclusive Economic Zones and the High Seas are protected from threats by the Year 2020.
> 3. Ameliorate or stop fishing methods that undermine sustainability by harming the habitats of economically valuable marine species and the species they use for food and shelter.
> 4. Stop physical alteration of terrestrial, freshwater and marine ecosystems that harms the sea, minimize pollution discharged at sea or entering the sea from the land, curtail introduction of alien marine species and prevent further atmospheric changes that threaten marine species and ecosystems.
> 5. Provide sufficient resources to encourage natural and social scientists to undertake marine conservation biology research needed to protect, restore, and sustainably use life in the sea.

Population Impact

Although population growth also has a big impact on the oceans, it's tough to change in the short term. Efforts aimed at reversing population growth and human impacts on marine systems include:

- Encouragement of sustainable use of ocean resources
- Policies and new technologies to limit pollution
- Creation and management of protected marine regions
- Education on ocean preservation and health and its human impacts

The oceans unite the people and land masses of the Earth. In fact, one in every six jobs in the United States is thought to be marine-related and attributed to fishing, transportation, recreation, and other industries in coastal areas. Ocean routes are important to national security and foreign trade. Military and commercial vessels travel the world on the oceans.

To highlight the world's oceans, the United Nations declared 1998 the International Year of the Ocean. This helped organizations and governments increase public awareness and understanding of marine environments and environmental issues.

❯ Review Questions

Multiple-Choice Questions

1. Groundwater is stored in

(A) porous rocks
(B) snow
(C) glaciers
(D) aquifers
(E) all the above

2. The layer closest to the surface, where spaces between soil particles are filled with both air and water, is called the zone of

(A) aeration
(B) hydration
(C) precipitation
(D) consolidation
(E) acclimation

3. Infiltration takes place when

(A) streams overflow their banks
(B) plants use water during photosynthesis
(C) deserts get drier and drier
(D) rainfall soaks into the ground
(E) snow melts

4. When water changes from a liquid to a gas or vapor, the process is called

(A) aeration
(B) conservation
(C) precipitation
(D) withdrawal
(E) evaporation

5. The study of the occurrence, distribution, and movement of water on, in, and above the surface of the Earth is called

(A) microbiology
(B) ecology
(C) hydrology
(D) botany
(E) aquaculture

6. A place where water is stored for some period of time (e.g. atmosphere, ocean, or underground) is called a

(A) dam
(B) stream
(C) lock
(D) reservoir
(E) diversion

7. The oceans hold approximately what percent of the Earth's water?

(A) 45%
(B) 62%
(C) 75%
(D) 80%
(E) 97%

8. Aquifers that form a water table separating unsaturated and saturated zones are called

(A) unconfined aquifers
(B) confined aquifers
(C) lakes
(D) closed aquifers
(E) saturated aquifers

9. The water found in subterranean spaces, cracks, and open pore spaces of minerals is called

(A) watershed
(B) groundwater
(C) snow melt
(D) runoff
(E) evapotranspiration

10. Rainfall or snow melt that has not had time to evaporate, transpire, or move into groundwater is known as

(A) transpiration
(B) an aquifer
(C) consumption
(D) runoff
(E) conservation

11. All the following are important water conservation methods except

(A) better farming techniques
(B) oscillating sprinkler systems
(C) dry cooling systems
(D) preventable runoff
(E) irrigation canals

12. The area at the surface or below ground that supplies water to an aquifer and/or most of the watershed or drainage basin is called

(A) an aquifer
(B) an unconsolidated aquifer
(C) a drainage zone
(D) an aquifer recharge zone
(E) a surface discharge

13. The amount of open space in the soil is called

(A) crystal structure
(B) sedimentation
(C) soil porosity
(D) granite
(E) soil salinity

14. Under surface soil, where all the open spaces have become filled with water, lies the zone of

(A) sedimentation
(B) saturation
(C) equalization
(D) salination
(E) diffusion

15. Pesticide and fertilizer use on crops affects water purity when

(A) evaporation is increased by heat
(B) it is used sparingly
(C) there is too little rainfall
(D) it is used in proper amounts
(E) runoff joins with surface and/or groundwater

16. The depletion of water from aquifers faster than they can naturally be refilled is called

(A) aquifer saturation
(B) groundwater mining
(C) flooding
(D) runoff
(E) the water table

17. The amount of withdrawn water lost in transport, evaporation, absorption, chemical change, or is otherwise unavailable as a result of human use is known as

(A) withdrawal
(B) conservation
(C) consumption
(D) discharge
(E) runoff

18. When water is blocked by a low-permeability material below the aquifer, it is known as

(A) an unconfined aquifer
(B) a perched water table
(C) a watershed
(D) a zone of aeration
(E) a diverted stream

› Answers and Explanations

1. E—Groundwater is stored in a variety of reservoirs for different amounts of time.

2. A—This zone, closest to the surface, contains gaps between soil particles filled with air or water.

3. D—Infiltration occurs when water molecules fill openings between soil particles.

4. E—Water goes from the liquid to the gaseous or vapor phase in evaporation.

5. C—The prefix *hydro* means "water" in Latin, (*–ology* means the "study of").

6. D—A reservoir holds collected water in its different forms: solid, liquid, and gas.

7. E—Nearly all (97%) of the Earth's total water is held in its oceans.

8. A—Aquifers form a water table; those separated into unsaturated/saturated zones are called unconfined aquifers since they flow right into the saturated zone.

9. B—Groundwater is defined as water found below the land's surface.

10. D—Water, affected by surface conditions and flow rate, needs time to evaporate or sink into the ground or it just runs off the land.

11. B—Oscillating sprinkler systems, which lose lots of water to evaporation, are not efficient.

12. D—When water enters an aquifer, it is said to recharge.

13. C—Loosely packed soils have lots of openings or pores between mineral grains. How tightly minerals are packed is known as porosity.

14. B—A material is saturated when no water or other substance can fill the open spaces.

15. E—Excess pesticides run off the land and may end up in streams or groundwater.

16. B—Extracting water before it can be replenished parallels the extraction of metals and mineral resources in mining.

17. C—When water withdrawal exceeds local supply, it is consumed or unusable for an extended time.

18. B—When a small, perched aquifer is disconnected from a larger aquifer below, it is separated by an unsaturated zone and a second, lower water table.

Free-Response Questions

1. Aquifers are often found in various places. Hydrologists are concerned about surface pollutant runoff making its way down into underground aquifers. Some aquifers form a water table that separates saturated and unsaturated zones. Some aquifers, however, lie beneath layers of impermeable materials like clay. Confined aquifers are more complex than unconfined aquifers that flow freely.

(a) Can groundwater pollution affect the integrity of aquifers? Explain.
(b) What is the difference between confined and unconfined aquifers?
(c) Where are aquifers generally found?

2. Use the chart in Figure 7.2 to answer the following questions.

(a) What activity uses the most water in the United States?
(b) What actions should cities take to conserve water resources?
(c) How can individuals reduce water use in their daily routines?

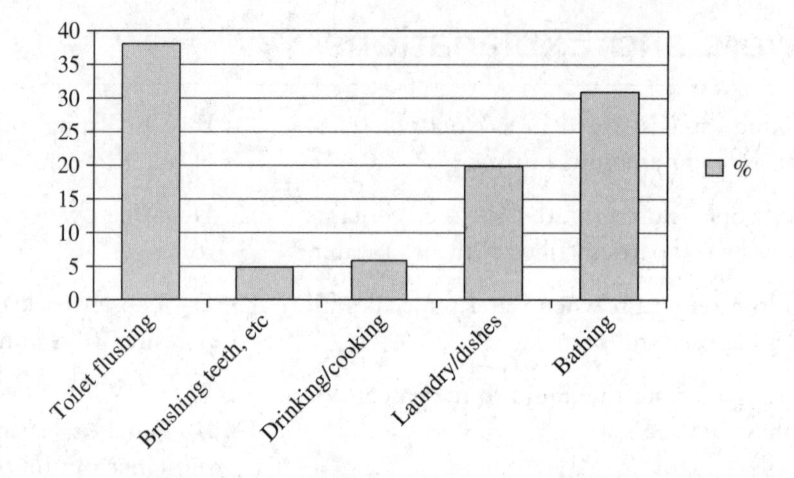

Figure 7.2 Percentage of water usage for an individual's activities.

Free-Response Answers and Explanations

1.

a. Yes. Because water flows with gravity and the path of least resistance; whatever is not soaked up by the ground finds its way elsewhere. In some cases polluted groundwater can make its way into an aquifer and dramatically affect the quality of the stored water. To be used, the water in the aquifer must undergo treatment at a high economic cost.

b. Unconfined aquifers are those that form a water table that separates the unsaturated and saturated zones since they flow right into the saturated zone. The water table of confined aquifers does not separate the two layers. Further, confined aquifers are most commonly under pressure causing water in some spots, like wells, to rise above the groundwater table.

c. Aquifers are generally found in areas where water has been redirected by an obstacle.

2.

a. Toilet flushing.

b. Mandate low-water toilets, scheduled lawn watering, and fines for water wastefulness, especially in the summer when aquifers are low.

c. Turn off running water when brushing teeth, doing dishes, and soaping up in the shower.

› Rapid Review

- Water shortages affect at least one-third of the world's population, with regional shortages being an issue.
- Aquifers are large underground rock formations that store water in reservoirs.
- Recharge zones allow water to enter aquifers.
- Water withdrawal refers to the total amount of water taken, while consumption refers to water lost to direct use, evaporation, and ground seepage.
- Runoff is made up of rainfall or snow melt that has not had time to evaporate, transpire, or move into groundwater locations.

- The geographical region from which a stream gets water is called a drainage basin or watershed.
- Evaporation is water to vapor; condensation is the reverse, vapor to liquid (water).
- Nearly all (97%) of the Earth's total water is contained in its oceans.
- Soil or rock is saturated when no water or other substance can fill the open spaces.
- Aquifers that form a water table that separates the unsaturated and saturated zones are called unconfined aquifers since they flow right into the saturated zone.
- Oscillating sprinkler systems lose a lot of water to evaporation and are not particularly efficient.
- The Gulf Stream current is about 800 times the volume of the Amazon, the world's largest river.
- Water table levels fall during times of low rainfall and high evapotranspiration.
- Aquifer recharge zones, either at the surface or below ground, supply water to an aquifer and/or most of the watershed or drainage basin.
- Porous media aquifers are made up of combined individual particles such as sand or gravel where groundwater is stored or moves between individual grains.
- The boundary between two watersheds is called a divide.
- Scientific concern about increasingly impacted oceans compelled more than 1,600 marine scientists to sign the 1998 *Troubled Waters* statement to raise public awareness.

Soil and Soil Dynamics

IN THIS CHAPTER

Summary: The Earth is made up of a core, mantle, and crust affected by volcanic, climatic, water, and weathering processes. The rock cycle describes the creation, destruction, and metamorphosis of rocks and minerals. Soil conservation maintains soil richness and crop vitality by protecting against erosion and desertification.

Keywords

✪ Sedimentary, igneous, metamorphic, rock cycle, stratigraphy, lithology, magma, tectonic, geothermal gradient, weathering, dissolution, hydrolysis, runoff, tillage, soil horizon, topography, desertification

Stratigraphic Classification

Unlike newly erupted volcanic rock, sedimentary rock strata or layers provide snapshots of regional climates and geological events throughout history. By studying different rock layers, geologists find windows into climate conditions during specific time periods.

Rock Stratigraphy

The study of rock, *stratigraphy,* is a grouping exercise. Figure 8.1 shows a cross section of the ancient (pre-Cambrian) and more recent (Paleozoic) sedimentary rock layers in the United States' Grand Canyon. Layers can be thin, thick, rocky, or smooth, but all have a place in the geological stack and time. The Grand Canyon is a good example of sedimentary rock layers above metamorphic and ancient rock.

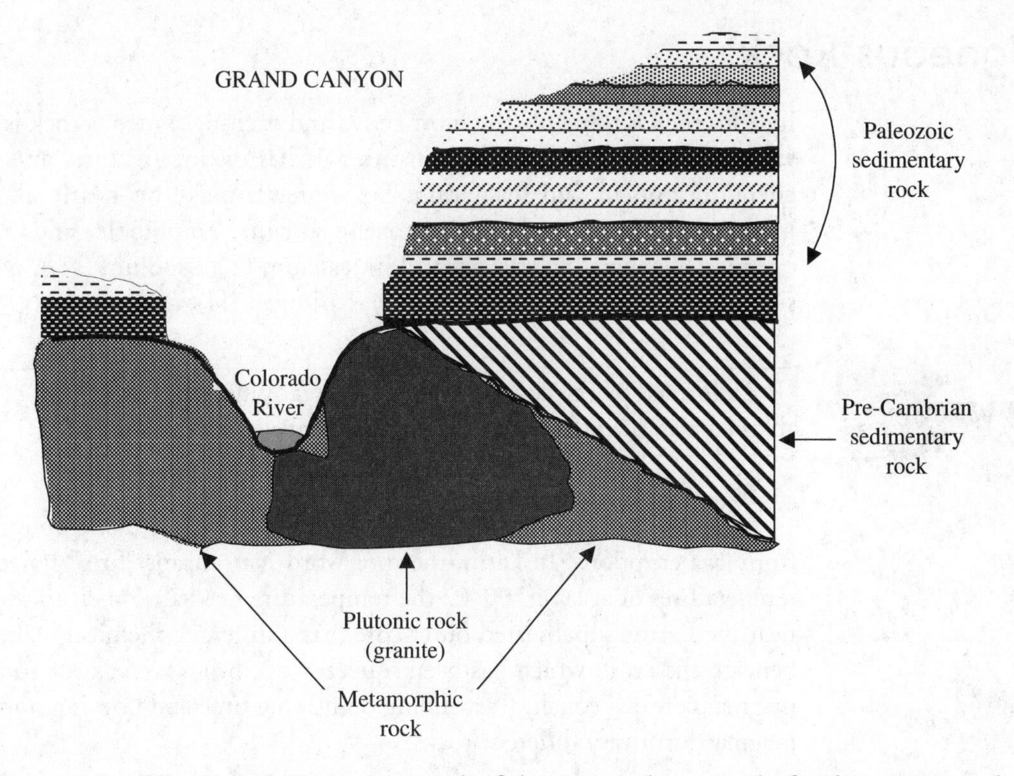

Figure 8.1 The Grand Canyon is a colorful stratigraphic record of sedimentary rock.

An individual band with its own specific characteristics and position is called a *rock–stratigraphic unit* or *rock unit.*

Rock units stacked vertically add up to a *formation,* which geologists can describe and map as part of the geological record. Formations are collections of many rock units grouped into a section with the same physical properties. Formations can be seen in places where strata layers are exposed. Igneous and metamorphic rock layers also have specific formations.

When studying sedimentary rock strata, it is important to remember huge stretches of time have led to layer upon layer of solidified rock. Thousands and millions of years piled atom upon atom, crystal upon crystal, building each layer.

Lithology is the visual study of a rock's physical characteristics using a handheld magnifying glass or a low-power microscope.

Different rock formations can be matched by their physical characteristics such as

- Grain size and shape
- Grain orientation
- Mineral content
- Sedimentary structures
- Color
- Weathering

The three main rock types are *igneous, sedimentary,* and *metamorphic.*

Igneous Rock

Igneous rock is probably the most active and exciting. Igneous rock is created by exploding volcanoes and boiling undersea fissures. It has distinct textures and colors depending on chemical content and formation. Six minerals make up nearly all igneous rock. These minerals are quartz, feldspar, pyroxene, olivine, amphibole, and mica. These minerals' chemical elements include silicon (Si), calcium (Ca), sodium (Na), potassium (K), magnesium (Mg), iron (Fe), aluminum (Al), hydrogen (H), and oxygen (O).

Rock formed by the cooling and hardening of molten rock (*magma*), deep in the earth or blasted out during an eruption, is called *igneous rock.*

Over 95% of the top 10 miles of the Earth's crust is made up of igneous rock formed from lava eruptions. In Latin, the root word *ignis* means "fire." Igneous rock is formed in temperatures of at least 700°C; the temperature needed to melt rock. The deepest magma, near the Earth's superheated outer core, has a different chemical makeup from magma just beneath the crust, which has been squeezed up through cracks or conduits. Not all cooled magma is created equally. Depending on heating time and how it got to the surface, different magmas form very different rocks.

Chemical and Mineral Composition

Igneous rocks are divided into two types depending on composition: *felsic* and *mafic.* Felsic rock is affected by heat, either from magma coming to the surface from extreme depths in the Earth, or by the friction between continental plates. Felsic rock has high levels of silica-containing minerals (e.g., quartz and granite). Mafic rock has high levels of magnesium and iron (ferric) minerals. The word *mafic* comes from a combination of these two mineral names.

Sedimentary Rock

Sediment is made up of loose particulate matter like clay, sand, gravel, and other bits and pieces of things. Sedimentary rock is formed when the bits and pieces of different rocks, soils, and organic things are compressed under pressure.

Sedimentary rock is formed from rocks and soils from other locations compressed with the remains of dead organisms.

Sedimentary rocks, originally formed from the buildup of material upon the Earth's surface, were compressed and cemented into solid rock over geological time. Beautiful multicolored layers of sedimentary rock are seen when highways or railroads cut across hillsides and mountains exposing the different types of sediments.

Rock Texture

In the late 1700s, while working in a field near his home in Scotland, James Hutton noticed coarse-grained granites cutting across and between layers of sedimentary rocks. Hutton thought physical changes in bordering sedimentary rock must have come from earlier contact with extreme heat. This gave him the idea that molten magma deep within the earth had squeezed into areas of sedimentary rock and crystallized.

KEY IDEA

Grain size and *color* are the two major ways geologists describe rock textures.

Mineral or crystal size, which makes up a rock's texture, is called *grain size.* Since color can change depending on lighting, mineral content, and other factors, it is considered less dependable when describing a specific rock.

When a rock's grains are easily seen with the eye (roughly a few millimeters across), they are classified as coarse-grained. When individual grains are not visible, the texture is reported as fine. Mineral grains or crystals have an assortment of different shapes and textures. They may be flat, parallel, needlelike, or equal in every direction like spheres or cubes. Granite has a coarse grain size compared to obsidian, which has a very fine grain size.

Lithification

Lithification comes from the Greek word *lithos,* meaning "stone." Lithified soil is made up of sand, silt, and organic organisms. Lithification can take place right after materials are deposited or much later. Compaction and cementation rate also play a big part in lithification. Additionally, the heat needed for lithification is less intense than that found deeper in the mantle, so lithification can occur in the top few kilometers of the crust.

Sandstone is formed when grains are squeezed together by the weight of overlying sediments during compaction, and formed into rock denser than the original sediments. These dense layers are then sealed by the precipitation of minerals in and among the layers.

Sediments become rock (lithified) through a process called *diagenesis.* Diagenesis is controlled by temperature. But instead of the hot igneous or metamorphic rock temperatures, diagenesis takes place at temperatures of around 200°C in sedimentary rock.

Diagenesis includes (1) compaction, (2) cementation, (3) recrystallization, and (4) chemical changes (like oxidation and reduction).

During diagenesis, unstable minerals recrystallize into a more stable matrix form or are chemically changed, like organic matter, into coal or hydrocarbons.

Compaction

Lithification through *compaction* is simple. As more and more sediments pile up, weight and pressure increase. The heavier the weight, the more the lower layers are compacted. The sediment's total volume is reduced since it is squeezed into a smaller space and dried. When shale grains are compacted, they align in the same direction, forming rock that splits along a flat plane in the same direction as the parallel grains.

Sediment *cementation* happens when compacted grains stick together. Minerals like calcite, silica, iron oxide, and magnesium cement into a dry, solid mass, becoming rock. Squashed sediments are so tightly ordered they shut out the flow of mineral-containing water and compact.

Sedimentary minerals can also be dissolved from flowing water creating pockets and places for other minerals to collect. Petroleum geologists look for oil in these pockets. When sedimentary minerals dissolve in water and form other compounds, it is called *dolomitization.*

Dolomitization happens when limestone turns into dolomite by a mineral substitution of magnesium carbonate for calcium carbonate.

Crystallization and Chemical Properties

Sedimentary rocks are classified by their chemical makeup and properties. Although silica (SiO_2) and phosphorus play a big part in the makeup of sedimentary rock, they are found in small amounts in seawater. When the water evaporates, the ions crystallize to form rock.

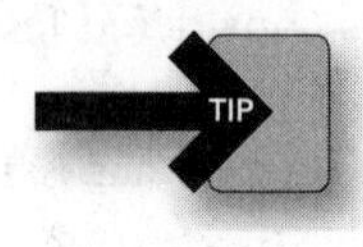

Carbonate sediments come from the biochemical precipitation of the decayed shells of marine organisms. Chemical sediments high in calcium (Ca^{2+}) and bicarbonate (HCO_{3-}) precipitate from seawater as calcium carbonate ($CaCO_3$) and carbonic acid (H_2CO_3) by inorganic processes.

Types of Sedimentary Rocks

Unlike igneous rock, most sedimentary rocks have a fine-grained texture. Layered or settled by water or wind, the sedimentary particles are usually small and fine.

Sedimentary rock deposition is also related to size. Fine silt grains are transported by wind, while water current tumbles rocks of larger sizes. The stronger the current, the farther it is carried. Rocks are grouped according to size when flowing in the same current stream.

Clastic

Clastic or *detrital* sedimentary rocks are formed when rocks have been carried to a different spot, weathered, and through pressure from overriding rock take a different form. They have a clastic (broken) texture made up of bigger pieces, like sand or gravel, and are grouped according to grain size. Table 8.1 lists the various clastic particles and their sizes.

Detritus is igneous, sedimentary, or metamorphic rock, which has been moved from its original location.

Clastic sedimentary rocks have particles ranging in size from microscopic clay to boulders. Their names are based on clast or grain size with the smallest grains being clay, silt, and sand. *Shale* is a rock made mostly of clay, *siltstone* is made up of silt-sized grains, *sandstone* is made up of sand-sized clasts, and a *conglomerate* is made up of pebbles surrounded by a covering of sand or mud.

Table 8.1 Different sediment types are found in a range of different particle sizes.

SIZE OF PARTICLE (MM)	SEDIMENT	ROCK
<1/256	Clay	Claystone or shale
1/256–1/16	Silt	Siltstone or shale
1/16–2	Sand	Sandstone
2–64	Pebble	Conglomerate or breccia
64–256	Cobble or gravel	Conglomerate or breccia
>256	Boulder	Conglomerate or breccia

Uniform Layer

A sedimentary rock layer made up of particles of about the same size is known as a *uniform layer.* A uniform clastic rock layer has single-sized particles tumbled by a constant-speed current. A uniform bed has single-sized particles, which have experienced different speed water currents and subsequently formed layers at different geological times.

Water, wind, and ice all work together to break down solid rocks into small rocky particles and fragments. These bits of rock are swept away by rain into streams. Gradually these particles get deposited at the bottom of stream beds or in the ocean. As more sediment builds up, it is crushed and compacted into solid rock.

Around 600 million years ago, minerals and microscopic organisms mixed with sediment to form layers. Extreme weight and pressure turned the ocean sediments into sedimentary rock.

Metamorphism

The term *metamorphism* comes from the Greek words, *meta* and *morph,* which mean "to change form." Geologists have found that nearly any rock can undergo metamorphism. When sedimentary and igneous rocks are exposed to extreme pressure and/or heat, they melt and form a denser, compacted *metamorphic rock* deep in the Earth's crust.

Metamorphic rocks are formed when rocks originally of one type are changed into a different type by heat and/or pressure.

Metamorphism changes a rock's original chemical and physical conditions. The three main forces responsible for the transformation of different rock types to metamorphic rock are the internal heat from the Earth, the weight of overlying rock, and the horizontal pressures from previously changed rock. Common metamorphic rocks include *slate, schist, gneiss,* and *marble.*

Rock Cycle

The deeper sedimentary layers are buried, the more the temperature rises. Extreme rock weight causes increased pressure and temperature. This heat and pressure cycle, describing the transformation of existing rock, is called the *rock cycle.* It is a constantly changing feedback system of rock formation and melting, which links sedimentary, igneous, and metamorphic rock. Figure 8.2 illustrates how the three rock types feed into a simple rock cycle.

During weathering, rock is either worn away or shifted from one spot to another. In the course of geological time, rock goes through an entire life cycle or rock cycle. One rock's lifetime might include being blasted out of a volcanic vent to the surface, settling back on the earth as a layer of volcanic ash, being lithified with other sediments into a sedimentary rock layer, and then being pushed down at a subduction zone to be transformed by pressure to metamorphic rock. Should the metamorphosed rock come in contact with a magma chamber or hot spot, it might melt and be shot to the surface. The cycle would be complete or could start all over again.

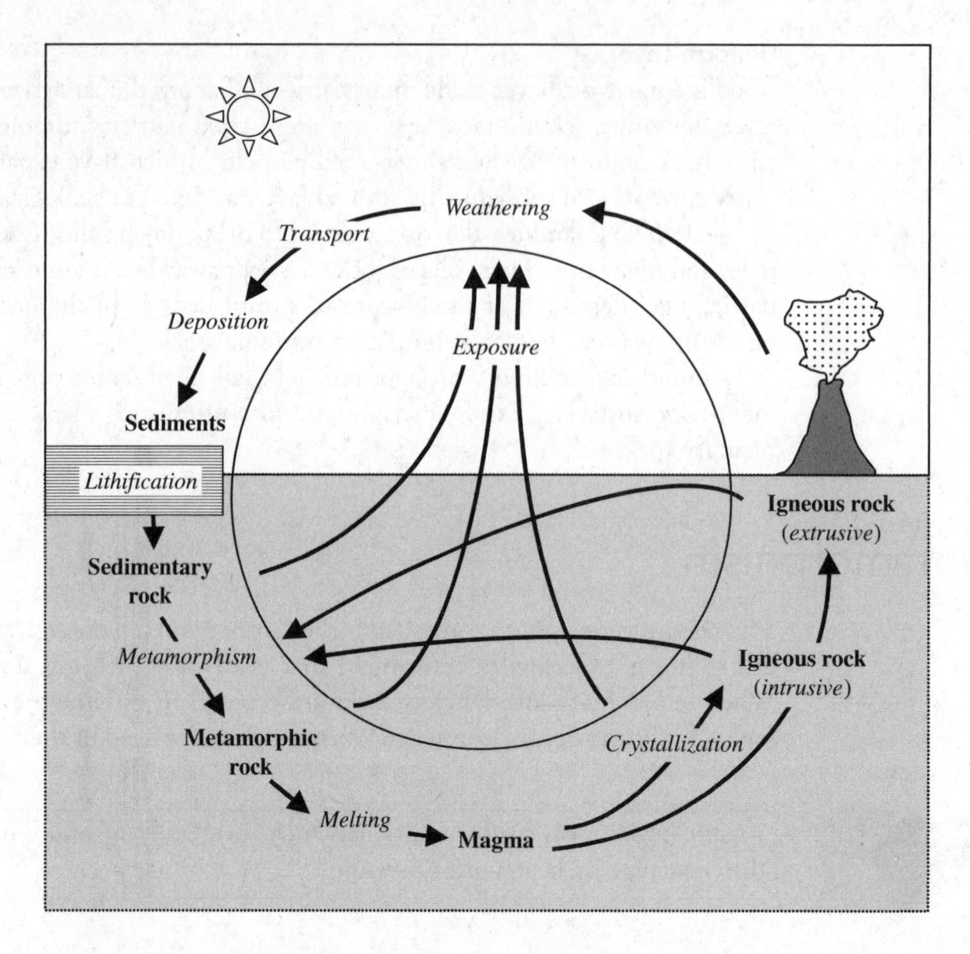

Figure 8.2 Weathering and transport of rock can follow many paths.

Like people, rocks are affected by their environment. We have learned how different rocks form; now let's look closer at the factors that play a part in aging.

Physical Weathering

The activities of physical and mechanical weathering create cracks in rock, which act as channels for air and water to get deeper into a rock's interior. During weathering, rock is constantly being broken into smaller pieces. The associated surface area, exposed to air and water, gets larger.

Physical weathering happens when rock gets broken (cracked, crumbled, or smashed) into smaller pieces without a change in its chemical composition.

Physical or mechanical weathering is the breakdown of large rocks into smaller bits that have the same chemical and mineralogical makeup as the original rock. Everyone is familiar with the breakdown of rock into smaller and smaller portions:

Boulders ⇒ pebbles ⇒ sand ⇒ silt ⇒ dust

This size-graded breakdown takes place in different ways. Table 8.2 shows the relationship between rock characteristics and the rate of weathering.

Table 8.2 Rocks weather at different rates depending on their characteristics.

ROCK CHARACTERISTICS	WEATHERING RATE		
Solubility	Low	Medium	High
Structure	Immense	Has weak points	Highly fractured
Rainfall	Low	Medium	High
Temperature	Cold	Moderate	Hot
Soil layer	No soil (bare rock)	Thin to medium	Thick
Organic activity	Negligible	Moderate	Plentiful
Exposure time	Brief	Moderate	Lengthy

Joints

Many rocks are not solid all the way through. They have many different size cracks and fractures, called *joints*, which are caused by stress.

Most jointing occurs in rock as a result of internal and external forces. These include:

- Cooling and shrinking of molten matter
- Flattening and tightening of drying sedimentary strata
- Plate tectonics

Joints are often arranged in sets of vertical parallel lines. The largest, most visible fractures are called *master joints*. Sometimes, multiple sets of joints intersect at nearly right angles and create a joint system made up of crossing vertical and horizontal joints.

In sedimentary rocks, various sets of joints often match planes separating strata. Along with joints, rocks can also fracture between individual crystals or grains. Weathering factors, like wind or water, enter tiny gaps between grains, leading to grain-by-grain disintegration of rocks.

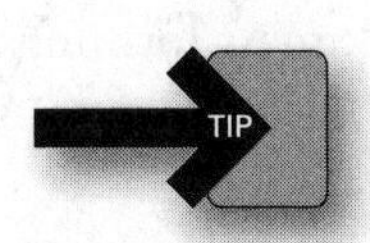

The two most important forms of physical rock breakdown are *joint block separation* and *granular disintegration.* Rocks break in several different ways. Some of these include:

- Frost
- Salt crystal growth
- Unloading (weight)
- Expansion and contraction due to the change of temperature and wetting-drying cycles
- Chemical weathering
- Biological weathering

Frost

Frost wedging is a type of mechanical weathering (i.e., weathering involving physical rather than chemical change). Frost wedging is caused by the repeated freeze–thaw cycle in extreme climates where water collects in the cracks or joints of rocks.

As the day cools and temperatures drop below freezing at night, the water inside the joints freezes and expands. Within rock fissures, expanding ice puts pressure on cracks until the pressure gets too high and the crack expands. Sometimes, rock splits the first time it expands, but often it takes many freezes and thaws until finally a joint succumbs and cracks.

Frost wedging happens when rock is forced apart by the alternate freezing and thawing of water.

Frost action is best seen in wet climates with many freezing and thawing cycles (arctic tundra, mountain peaks). Frost splits rocks into blocks and wears away block edges grain by grain, rounding the surfaces.

Salt Wedging

Most people think of deserts as barren, dry, windy, inhospitable landscapes with circling vultures to keep you company. Everything in a desert environment is *desiccated* (dried out) much of the time and subject to weathering.

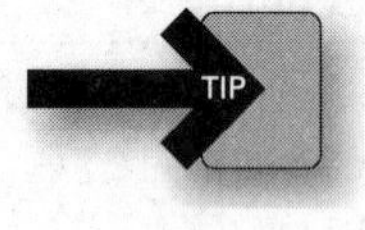

Salt wedging, caused by salt crystal growth, is an important rock-breaking force in the desert. In dry and semi-desert environments, surface water and soil moisture evaporate quickly. When this happens, dissolved salts fall out of solution and crystallize. Growing salt crystals, such as halite ($NaCl$), calcite ($CaCO_4$), and gypsum ($CaSO_4$), put pressure on the rock. Over time, this pressure wears bedrock away grain by grain.

Unloading

Rocks buried within the crust are under extreme pressure from the weight of overlying rock. Over time, erosion and mass wasting remove upper rock layers, and the pressure on lower rock layers becomes less. When this happens, the internal rock volume expands and the outer rock layers crack away. This shedding of outer layers from internal pressure changes is called *unloading*. It's like a snake that sheds its skin as it grows larger.

Unloading happens when there is a release of internal rock pressure from erosion and outer rock layers crack off.

When rock is worn away on the surface, it is called *exfoliation*. Exfoliation occurs when large sheets of rock expand upward and shear away from the primary rock mass forming a dome. The majestic Half Dome in Yosemite Valley, California, is an exfoliation dome.

Chemical Weathering

When mechanical weathering breaks a rock apart, the larger exposed surface allows *chemical weathering* to take place. These chemical and structural changes happen over millions of years. However, human-made industrial pollutants in the air and water have caused some forms of chemical weathering to increase.

When rock and its component minerals are broken down or altered by chemical change, it is known as *chemical weathering*.

Chemical weathering takes place in one of the following ways:

- *Oxidation* = reaction with O_2
- *Hydrolysis* = reaction with H_2O
- *Acid action* = reaction with acid substances (H_2CO_3, H_2CO_4, H_2SO_3)

The most important natural acid is *carbonic acid,* formed when carbon dioxide dissolves in water ($CO_2 + H_2O \leftrightarrow H_2CO_3$). Carbonate sedimentary rocks, like limestone and marble, are extra sensitive to this type of chemical weathering. Gouges and grooves seen in carbonate rock outcrops are examples of chemical weathering.

Acid rain is the result of a chemical reaction between atmospheric chemicals and water. When it falls to Earth, it speeds chemical weathering through a process called *dissolution.* When acid rain comes into contact with limestone, monuments, and gravestones, it dissolves, discolors, and/or disfigures the surface by reacting with the rock's composition. Historical statues and buildings, hundreds to thousands of years old, suffer from chemical weathering, a by-product of industrial pollution.

- *Oxidation* takes place when oxygen anions react with mineral cations to break down and form oxides, such as iron oxide (Fe_2O_3), which softens the original element.
- *Solubility* is the ability of a mineral to dissolve in water. Some minerals dissolve easily in pure water. Others are even more soluble in acidic water. Rainwater combined with carbon dioxide forms carbonic acid ($CO_2 + H_2O = H_2CO_3$) becoming naturally acidic.
- *Hydrolysis* takes place when a water molecule and a mineral react to form a new mineral.
- *Dissolution* occurs when environmental acids like carbonic acid (water), humic acid (soil), and sulfuric acid (acid rain) react with and dissolve mineral anions and cations.

Biological Weathering

Biological weathering is a blend of both physical and chemical weathering. Tree and plant roots grow into rock fissures to reach collected soil and moisture. As they grow, roots get thicker and push deeper into the crack. Eventually, this constant pressure cracks the rock. The more rock is fractured, the easier it cracks. Root growth and burrowing in rock gaps cause rock fractures to open wider and become more exposed to future weathering.

In biological weathering, element and nutrient exchange occurs. Bacteria and algae in cracks and on rock surfaces have acidic processes, which add to chemical weathering. Plants get needed minerals from rock and soil.

Soil Erosion

Erosion converts soil into sediment. Chemical weathering produces clays on which vegetation can grow. A mixture of dead vegetation and clay creates soil that contains minerals plants need for growth.

Soil exists as a layer of broken, unconsolidated rock fragments created over hard, bedrock surfaces by weathering action.

Most geologists classify soil into three layers called *soil horizons* or *soil zones.* These soil horizons are commonly recognized as horizons A, B, and C, but not all three horizons are found in all soils. Figure 8.3 illustrates the way soil horizons are stacked on top of each other.

Soil horizons are described from the top soil layer down to the lowest soil and bedrock level and are as follows:

- *Horizon A* includes the surface horizon, a zone of leaching and oxidation, where penetrating rainwater dissolves minerals and carries the ions to deeper horizons. It also holds the greatest amount of organic matter.

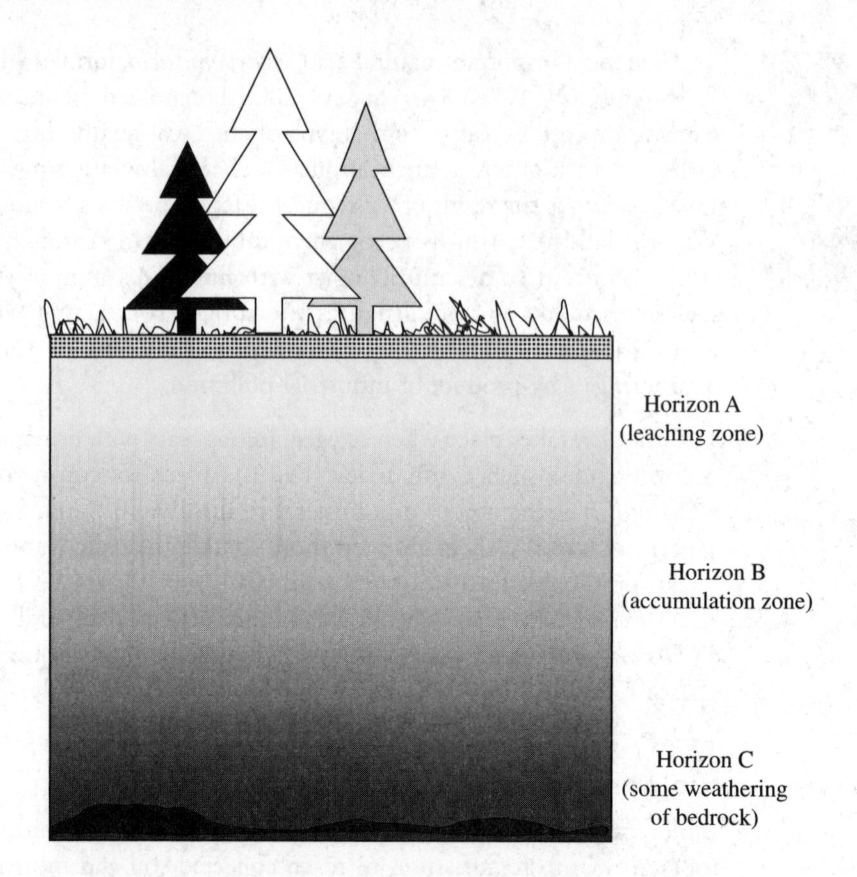

Figure 8.3 Rock can be divided into a gradient of soil horizons with bedrock at the bottom.

- *Horizon B* describes the middle horizon, a zone of accumulation, where ions carried down by infiltrating rainwater are reconnected to create new minerals. Blocky in texture, it is made up of weathered rock mixed with clay, iron, and/or aluminum.
- *Horizon C* includes the bottom horizon, which is a zone of unconsolidated, weathered original rock.

Soil Types

Just as there are three different soil horizons, there are also several factors that affect soil formation. These include structure, rainfall (lots or little), solubility, temperature (hot or cold), slope (gentle or steep), vegetation (types and amount), and weathering time (short or long). Singly or in combination, soils form as a result of many factors. A key factor in naming major soil types is rainfall amount. Everyone from toddlers making mud pies to petroleum geologists looking for oil can tell whether a soil is wet or dry, hard or soft.

Geologists have named three basic soil types based primarily on water content. These are the *pedocal, pedalfer,* and *laterite.*

Pedocal, found in dry or semi-arid climates with little organic matter, has little to no mineral leaching and is high in lime. Most nutrient ions are still present. In places where water evaporates and calcite precipitates in horizon B, a hard layer called the *caliche* or *hardpan* is formed. Pedocal soil is found in dry climates with little rainfall. It supports mostly prairie plant growth.

Enriched with aluminum and iron, pedalfer soil is found in wetter environments and contains greater amounts of organic matter and leaching. Pedalfer is found in high temperatures and humid climates with lots of forest cover.

Laterite, the soggiest soil type, is found in tropical and subtropical climates and is high in organic matter. Because of heavy rainfall, there is widespread nutrient leaching of iron and aluminum hydroxides, which make laterite soils red.

Soil Conservation

Topsoil is the nutrient-rich soil layer, millimeters to meters deep, that contains a mixture of organic material and minerals. It is a renewable resource when replenished and cared for properly, but unfortunately billions of acres of land worldwide are bare due to erosion, nutrient depletion, overtillage, and misuse. It has been estimated that over 25 billion tons of soil are lost from cropland annually from wind and water erosion. In fact, roughly one-third of the world's land is at risk of *desertification.* However, there are several soil conservation methods that can help with sustainability including land management, soil enhancement, ground cover, and agricultural tillage methods.

Vegetation

Topsoil can be blown away by the wind or washed away by rainfall increasing erosion. Weakened land may also allow downstream flooding, reduced water quality, increased river and lake sedimentation, and the buildup of silt in reservoirs and navigation channels. It can also be a source of dust storms and air pollution. Wind-blown dust can increase health problems, including allergies, eye infections, and upper respiratory problems.

Topography

Since water runs downhill, it is easy for soil to be carried away during a heavy rainstorm. Slope percentage affects the speed of water's downhill path (e.g., a 5% grade contributes more to soil erosion than a 1% grade).

Topography is the mapping of the land contours and physical features of an area.

Farmers have also found that by changing planting and harvesting methods, they protect their land's fertility. *Contour planting* across a hillside instead of up and down slows runoff. When combined with *strip farming,* planting alternating crops in strips across land contours, erosion is slowed further. With one crop in the field holding the soil, the other crop is harvested. Alternating rows hold water longer and let rainfall soak into the ground avoiding runoff.

Terracing is a lot like strip farming, but the land is shaped as well. Level ridges of land are created to hold water and soil in place. Although more expensive and time consuming at first, it allows cultivation on steep grades, while increasing sustainability. In Asia, rice has been cultivated in this way for centuries, taking advantage of all fertile land.

Planting and Tillage

Planting perennial plants is another tool in soil conservation. Plants like coffee and tea, which grow during several seasons, don't have to be harvested yearly and hold the soil longer. Ground cover crops like clover or alfalfa, planted right after initial harvest, also hold and protect the soil from erosion.

Leaving crop remnants in the fields and adding *mulch* cover (e.g., wood chips, manure, straw, leaves, and decomposed organic matter) to crops after harvest holds the soil and adds nutrients.

Plowing or *tillage* often increases soil erosion. Thought to increase nutrients, broad plow tillage was done in the United States for decades. Today, narrow chisel plows leave 75% of crop residue on the surface and open only a thin ridge for seeds. No-till methods pierce seeds through ground cover without opening up a seam in the earth. This keeps soil in place and prevents erosion.

› Review Questions

Multiple-Choice Questions

1. Lithification, comes from the Greek word *lithos*, meaning

(A) stone
(B) light
(C) paper
(D) seawater
(E) white

2. Diagenesis causes

(A) a red rash on the elbow
(B) miscalculations of circular core samples
(C) photosynthesis in most upper ocean layers
(D) lithification of sediment by physical and chemical processes
(E) drought in some areas

3. Which of the following is a common metamorphic rock type?

(A) Subjugate rock
(B) Slate
(C) Vermiculite
(D) Slant
(E) Magma

4. Metamorphic rock is known as a

(A) hard purple-colored rock
(B) surface-only rock
(C) chameleon of rock types
(D) good building material
(E) brittle rock type

5. When several rock–stratigraphic units are stacked vertically, they are called a

(A) metamorphic shelf
(B) gneiss
(C) lithification
(D) vertical unit
(E) formation

6. When rock on the Earth's surface is worn away, it is called

(A) foliation
(B) metamorphism
(C) exfoliation
(D) sedimentation
(E) coloration

7. Clastic sedimentary rocks are made up of

(A) pieces of other rocks
(B) gelatin from kelp
(C) the hardest granites
(D) boulders larger than 1 meter across
(E) petroleum distillates

8. Which of the following plays a big part in soil erosion?

(A) Fertilizer
(B) Conservation
(C) Tillage
(D) Plant type
(E) Rainfall

9. Which process wears away existing rocks and produces lots of small rock?

(A) Lithification
(B) Sedimentation
(C) Compaction
(D) Weathering
(E) Concretion

10. Salt wedging, caused by the growth of salt crystals, is an important rock-breaking force in the

(A) rain forest
(B) mountains
(C) ocean
(D) plains
(E) desert

11. The following are all tools in soil conservation except for

(A) cover crops
(B) deforestation
(C) planting of perennial plants
(D) adding mulch
(E) crop rotation

12. Which metamorphic rock is often used as a paving and building stone?

(A) Mica
(B) Marble
(C) Shale
(D) Sandstone
(E) Obsidian

13. Felsic and mafic are two types of

(A) metamorphic rock
(B) clastic formations
(C) sedimentary rock
(D) igneous rock
(E) multicolored clays

14. What percentage of the world's land is at risk of desertification?

(A) 10%
(B) 20%
(C) 30%
(D) 40%
(E) 50%

15. Pebbles surrounded by a covering of sand or mud is known as

(A) a conglomerate
(B) shale
(C) granite
(D) magma
(E) igneous clay

16. The description of an area's land contours and surface features is called

(A) geography
(B) cartography
(C) graphology
(D) topography
(E) photography

17. Pedocal, pedalfer, and laterite are three basic soil types based primarily on

(A) elevation
(B) soil type
(C) water content
(D) mineral base
(E) climate

18. Which soil horizon consists of a zone of accumulation, where ions transported by rainwater are reconnected to create new minerals?

(A) Horizon A
(B) Horizon B
(C) Horizon C
(D) Ions are not transported this way
(E) Mineralization occurs only in the topsoil

› Answers and Explanations

1. A

2. D—Diagenesis is controlled by temperature and takes place at temperatures around 200°C in sedimentary rock.

3. B—Other metamorphic rock types include schist, gneiss, and marble.

4. C—Internal heat from the Earth, overlying rock weight, and horizontal pressures from previously changed rock cause metamorphic rock changes.

5. E—This layering can easily be seen in the Grand Canyon.

6. C—It describes how atoms and layers are peeled away very slowly.

7. A—These are formed from rocks carried from a different spot, weathered, and turned into new rocks.

8. C—Breaking up large amounts of soil by tilling the land allows particles to erode.

9. D

10. E—When water evaporates, salt remains and slowly increases.

11. B—Removing trees often involves disturbing the soil as well.

12. B

13. D—Felsic rock has high levels of silica, and mafic rock has high levels of magnesium and iron (ferric) minerals.

14. C

15. A—This type has a mixed bumpy texture.

16. D—A topographical map shows specific features like rivers, canyons, and mountains.

17. C—They are found in various climates with different rainfall amounts.

18. B—Horizon A (surface horizon) is a zone of leaching and oxidation, while horizon C (bottom horizon) is a zone of unconsolidated, weathered original rock.

Free-Response Questions

1. A heat and pressure gradient promotes metamorphism in a graded way depending on depth. The deeper you go, the hotter the temperature and pressure, the greater the metamorphic changes. Depending on the conditions under which rock is changed, the rock gradient forms either high-grade or low-grade metamorphic rock.

↑ Temperature and ↑ pressure ⇒ high-grade metamorphic rock
↓ Temperature and ↓ pressure ⇒ low-grade metamorphic rock

Rock's crystalline structure changes as it adjusts to new temperatures and pressures. Ions are energized, breaking their chemical bonds and creating new mineral linkages and forms. Sometimes, crystals grow larger than those in the original rock. New minerals are created by the rearrangement of chemical bonds or reactions with fluids entering the rock.

(a) What role does metamorphism play in the rock cycle?
(b) Explain the correlation between pressure and depth.
(c) Are greater metamorphic changes likelier to occur at a depth of 170 km or 1,700 km in the Earth's crust?
(d) Give an example of each of the following rock types: sedimentary, igneous, and metamorphic. How do they differ?

Free-Response Answers and Explanations

1.

a. Metamorphism is one process by which rocks undergo transition within the rock cycle. It occurs when rocks undergo changes due to the application of heat and pressure. Metamorphic rocks are created when igneous or sedimentary rocks undergo recrystallization due to extreme temperature and pressure changes.

b. Greater depths produce greater pressure due to the gradual accumulation of weight at greater depths.

c. Changes are more likely at 1,700 km due to greater temperature and pressure.

d. Sedimentary: sandstone; metamorphic: slate; igneous: sand. The effects of heat and pressure from the three processes in the rock cycle cause differences in color and texture.

❯ Rapid Review

- Topography is the mapping of an area's land contours and physical features.
- When rock units are stacked vertically, they create a formation in the geological record.
- The three main rock types are sedimentary, igneous, and metamorphic.
- The pedocal is found in dry or semi-arid climates where there is little organic matter, little to no leaching of minerals, and a high lime content.
- Lithology is the study of the physical characteristics of a rock through visual recording or with a low-power microscope or handheld magnifying glass.
- Lithified soil is made up of sand, silt, and organic organisms.
- Physical weathering occurs when rock gets broken into smaller pieces without a change in its chemical composition.
- Sedimentary rock is formed from rocks and soils from other locations that are compressed with the remains of dead organisms.
- Metamorphic rocks are formed when rocks originally of one type are changed into a different type by heat and/or pressure.
- Diagenesis includes (1) compaction, (2) cementation, (3) recrystallization, and (4) chemical changes (e.g., oxidation/reduction).
- Igneous rock is formed by the cooling and hardening of molten rock (magma), deep in the Earth or blasted out during an eruption.
- Igneous rock is divided into two main types: felsic and mafic.
- Felsic rock has high levels of silica-containing minerals (e.g., quartz and granite).
- Mafic rock has high levels of magnesium and iron (ferric) minerals.
- An individual band with its own specific characteristics and position is called a rock–stratigraphic unit or rock unit.
- Grain size and color are the two main ways that geologists describe rock textures.
- Acid rain is a chemical reaction that speeds up chemical weathering.
- The most important natural acid is carbonic acid, formed when carbon dioxide dissolves in water ($CO_2 + H_2O \leftrightarrow H_2CO_3$). It is a big part of acid rain.
- Frost wedging happens when rock is pushed apart by the alternate freezing and thawing of water in cracks.
- Salt wedging, caused by salt crystal growth, is an important rock-breaking force in the desert.
- Unloading happens when there is a release of internal rock pressure from erosion and the outer layers of a rock are shed.
- Turning soil over by the plow or tillage often increases soil erosion.
- Dolomitization happens when limestone turns into dolomite by a mineral substitution of magnesium carbonate for calcium carbonate.

CHAPTER 9

Ecosystem Structure, Diversity, and Change

IN THIS CHAPTER

Summary: The world contains millions of diverse plant, animal, and insect organisms living together in complex interconnected ecosystems. A major change in one group has a ripple effect on the overall system, sometimes impacting thousands of species.

Keywords

✪ Ecosystem, biosphere, biome, sustainable use, endemic species, range, extinct, wetlands, hotspots, ecological niche, habitat, primary succession, secondary succession, gene pool, natural selection, adaptation

Biological Populations

The world's oceans make up 99% of the planet's biosphere and contain the greatest diversity of life. Even the most biologically rich tropical rain forests can't match the biodiversity (number of species) found in a coral reef community.

Rain forests, deserts, coral reefs, grasslands, and a rotting log are examples of ecosystems with specialized populations. Land-based ecosystems are known as *biomes* and are further classified by rainfall and climate. Table 9.1 lists different biomes and their characteristics. Marine or aquatic-based ecosystems are primarily described as freshwater or saltwater. The vertical depths and sunlight levels are illustrated for an ocean biome in Figure 9.1.

An *ecosystem* is a complex community of plants, animals, and microorganisms linked by energy and nutrient flows that interact together and with their environment.

Table 9.1 The Earth's biomes have a variety of plant types and rainfall amounts.

BIOME	YEARLY RAINFALL/SOIL TYPE	PLANT LIFE
Tundra	<25 cm, permafrost soil	Small leafy plants
Deserts (hot and cold)	<25 cm, sandy, coarse soil	Cactus and other water storing plants
Chaparral (scrub forest)	50–75 cm (winter), shallow infertile soil	Small hard-leaved trees, scraggly shrubs
Conifer forest	20–60 cm, acidic soil	Waxy, needle-leaved trees (conifers)
Grasslands	10–60 cm, rich soil	Mat-forming grasses
Deciduous forest	70–250 cm, high organic composition soil	Hardwood trees
Tropical rainforest	200–400+ cm, low organic composition soil	Tall trees with associated vines, etc., adapted to low light

Because the oceans seemed limitless, it's hard to grasp pollution's heavy impact on plant and animal marine species and ecosystems. Within the last 30 years, population increases, new technology, increased seafood demand, and many other factors have impacted marine ecosystems in ways unknown 100 years ago. As the planet's population has passed seven billion, scientists, economists, policy makers, and the public are becoming increasingly aware of the strain on the oceans' natural ecosystems and resources.

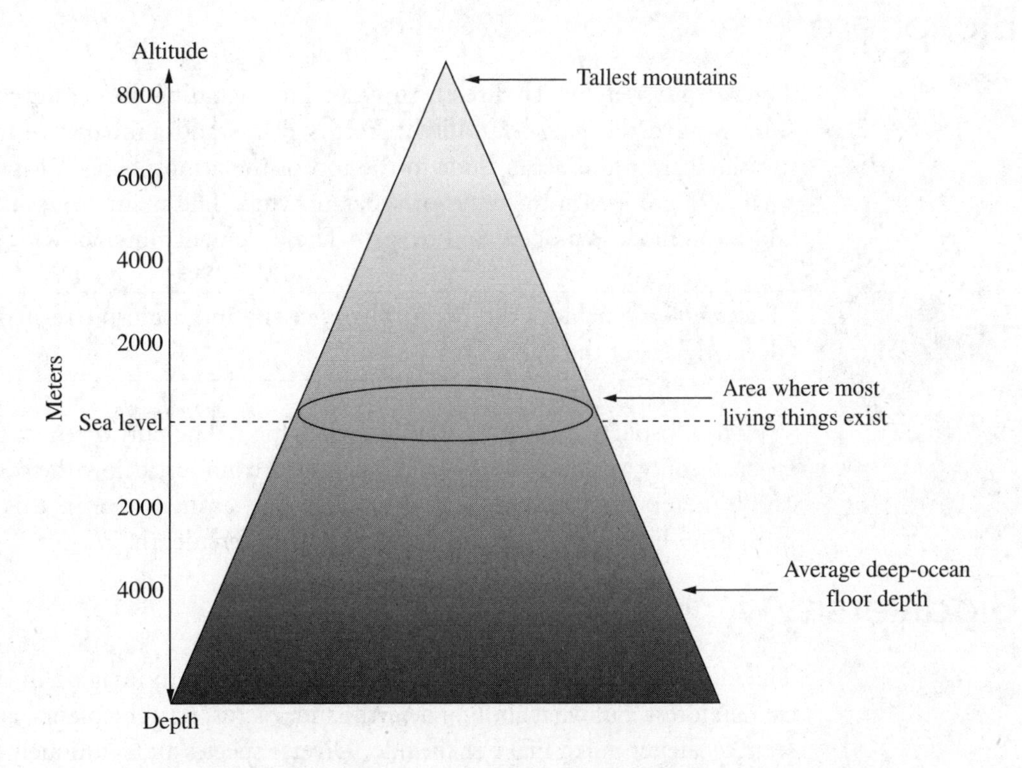

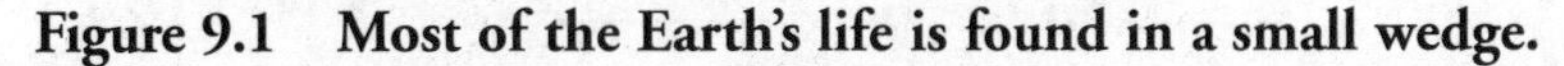
Figure 9.1 Most of the Earth's life is found in a small wedge.

Climate Change

The Intergovernmental Panel on Climate Change (IPCC) has gathered hundreds of scientists to examine and report on the record-setting rise in global temperatures, hazards, impacts, risk, and vulnerability of species and ecosystems. In 2014, a worldwide report was released with data showing impacts from climate-related extremes such as droughts, floods, hurricanes, and wildfires. Reported impacts for countries at all levels of development, included altered ecosystems, disrupted food production and water supply, damaged settlements, disease, mortality, and decreased human well being. From preindustrial times, carbon dioxide (280 ppm to 395 ppm in 2014), methane (718 ppb to 1893 ppb in 2014), and nitrous oxide (270 ppb to 326 ppb in 2014), have risen significantly. They act by absorbing infrared heat radiating from the Earth and heating the lower atmosphere. This atmospheric warming adds to natural greenhouse heating and causes even higher temperatures.

Changing climate temperatures impact coral reefs and forest ecosystems, along with related industries and jobs (lumber and fishing). Public policy in many countries has begun to address climate issues at the national, regional, and international levels. Conservation and sustainable biodiversity activities are becoming more common with a strong focus directed toward sustainable use.

Sustainable use is the use of resources in a way that protects the numbers and complexity of a species or environment without causing long-term loss.

Some of the biologically diverse areas currently under study include marine, coastal, and inland waters; and island, forest, agricultural, as well as dry, subhumid, and mountain regions. Research programs addressing basic principles, key issues, potential output, timetables, and future goals of single and overlapping systems are being created.

Biosphere

The Earth system, which directly supports life, including the oceans, atmosphere, land, and soil, is called the *biosphere.* All the Earth's plants and animals live in this layer, which is measured from the ocean floor to the top of the atmosphere. All living things, large and small, are grouped into *species* or separate types. The main compounds of the biosphere contain carbon, hydrogen, and oxygen. These elements interact with other Earth systems.

The *biosphere* includes the hydrosphere, crust, and atmosphere. It is located above the deeper layers of the Earth.

The biosphere is roughly 20,000 meters high. The portion most populated with living species is only a fraction of that. It is measured from just below the ocean's surface to about 1,000 meters above it. Most living plants and animals live in this narrow layer of the biosphere. Figure 9.1 gives an idea of the biosphere depth.

Biodiversity

The idea of a biologically diverse environment is easy to imagine in the middle of a tropical rain forest, but what about a desert? Sand, cactus, scrubby plants, and stunted trees don't seem to shelter much life, but they do. Diverse species are as uniquely suited to desert environments as those in the rain forest.

Biodiversity is a measure of the number of different individuals, species, and ecosystems within an environment.

An animal or plant with a specific relationship to its habitat or other species, filled by it alone, occupies an *ecological niche.* Interrelationships between ecological niches make up a complex ecosystem. Whenever a major species overlap exists or a foreign species is introduced, the local balance is upset. A new natural balance must be gained for the ecosystem to work smoothly again.

If biodiversity is unbalanced and species eliminated, then niches must adjust. Some adjustments are minor, but more often a domino effect takes place with all members of the ecosystem rebalancing. The groups that can't change die out.

Species Evolution

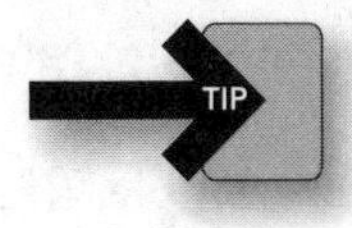

A population's total genetic makeup is called its *gene pool.* When climate or habitat allows some individuals to live and reproduce while others die, the population is undergoing *natural selection.*

Species reach levels of specialization through *adaptation* (i.e., changing as the environment changes) and *evolution* (i.e., species' gene pools change over time). Organisms that adapt and reproduce in changing environments are said to have *evolutionary fitness.*

Genetic material in a cell's nucleus often changes during evolution. This comes from random protein breaks or other changes known as *genetic drift.* When a large percentage of a population dies from disease, starvation, or predation, the population's remaining genetic material is much reduced. The frequency of certain traits is also narrowed from the original population causing a *bottleneck effect.*

A different type of genetic drift happens when a few members of a population migrate and create a new population. Again, only a fraction of the original genetic material and its diversity is represented in the new group. This is known as a *founder effect.* When a species is transported to a completely new region, it is known as an *alien species.* Alien species often compete with local species for resources.

The *Hardy-Weinberg principle* states that in a stable population, the frequency of genotypes and alleles (parts of genetic material) will remain constant.

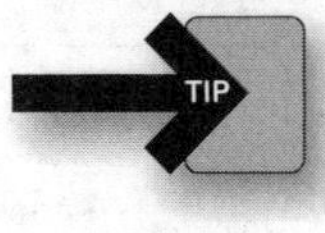

A *species* is a group of intrabreeding organisms unable to interbreed with a different species. Over geological time, two groups of the same species may change (*macroevolution*) and eventually become so different they can no longer reproduce with each other. This is an example of *allopatric speciation* and often results from reproductive isolation. This happens in animal and plant species. *Sympatric speciation,* or the evolution of two different species from a single species, is more common in plants.

Species Interaction

Nearly all species are interconnected within an ecosystem. *Interspecific competition* occurs when two or more different species need the same habitat or resources to survive. When members of the same species need the same resources (e.g., finding a mate), it is known as *intraspecific competition.*

When two species compete and the stronger, better adapted species wins, the process is called *competitive exclusion.* When a species has no competitive limitations, it has a *fundamental niche.* When a species must settle for a smaller niche than it normally would have because of competition, the niche is called a *realized niche.*

Gause's principle explains that no two species can fill the same niche at the same time, and the weaker species will fill a smaller niche, relocate, or die off.

Another type of interspecies competition, *predation,* occurs when one species serves as food for another species (e.g., rabbits for eagles). *Symbiosis* is the close, extended relationship between organisms of different species that may (↑) or may not (↓) benefit each participant. There are three types of symbiotic relationships to remember:

- *Mutualism.* Both species benefit (↑↑). (For example, lichen is a combination of a fungus and a photosynthetic algae or cyanobacterium. One adds structure and stores water, while the other creates organic compounds by photosynthesis.)
- *Commensalism.* One species is fairly unaffected (~) and one species benefits (↑). (For example, water buffalo and egrets have a commensalistic relationship because the egret eats insects from the buffalo's hide.)
- *Parasitism.* One species benefits (↑) and one species is harmed (↓). (For example, fleas are a parasite that live on dogs.)

Camouflage and *mimicry* are related competitive mechanisms. Many northern mammals (e.g., arctic hares) have coats that turn white in winter to blend with the snow, giving them protection or camouflage from predators. A walking stick is a brown, thin insect that looks just like a twig when clinging to a branch. It mimics its surroundings for protection.

Endemic Species

Plants and animals adapted to their environment are found worldwide. Some are widespread, while others are only found in a single river, lake, island, or mountain range. Organisms unique to a specific area are called *endemic species.*

Species like dogs and cats live in many habitats. Even in the wild, they are widely scattered. However, most species are limited to certain areas because their ecological requirements are only found in a small area. They might flourish in another region, but they aren't able to travel long distances or cross deserts to get there.

Endemic species, naturally occurring in only one area or region, are unique to that specific region.

Polar bears aren't found in Arizona, because they are endemic to polar regions. Plants and animals needing warmer climates or a longer growing season are restricted by environmental conditions like temperature and rainfall. A species' geographical *range* often stretches across wide areas, depending on the environmental conditions. When their core habitat needs are met, they survive.

The entire area that a plant, animal, insect, or other organism travels during its lifetime is considered its *range.*

The range of the once limitless American bison (millions of animals) has been reduced to a tiny fraction of what it once was. Range loss, and the massive slaughter that took place during the 19th century construction of the North American east–west railroad across their territory, took a heavy toll on the bison.

Keystone Species

A species around which an entire ecosystem is dependent, is known as a *keystone* or foundational species. Its controlling influence protects and balances many other species. For example, conservationists reintroduced wolves into Yellowstone Park in Wyoming to balance the number of elk and other species, since without the predator they were outstretching regional resources.

Ecological Succession

Biological communities don't just suddenly spring up fully developed; there is a sequence of development. Over time, a previously untouched area like a volcanic flow will be populated by different species creating soil, cover, shade, and food resulting in a *primary succession* of development.

The first colonizers to a site (e.g., moss or lichens) are called *pioneer species.*

Secondary succession takes place when an existing community is disrupted by some event (e.g., wildfire, mining, or plowing) and must begin again. In both cases, the process of environmental modification by biological organisms is called *ecological development.* Less developed species gradually give way to more developed species as in the following succession example.

Rock ➡ lichen ➡ moss ➡ grasses ➡ bushes ➡ seedlings ➡ pine trees ➡ oak trees

When an ecosystem reaches its final stage of balanced species development, it is known as a *climax community.*

Habitat

The area in which an animal, plant, or microorganism lives and finds nutrients, water, sunlight, shelter, living space, and other essentials is called its *habitat.* Habitat loss, which includes the destruction, degradation, and fragmentation of habitats, is the primary cause of biodiversity loss.

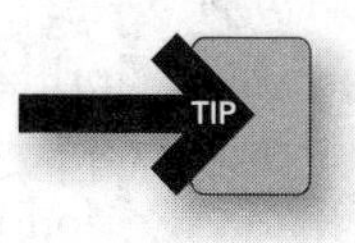

Loss of habitat is perhaps the most important factor affecting a species. Think of when a tornado or hurricane levels a town. Not only are homes and businesses lost, but water supplies, food crops, communications, and transportation may also be annihilated. The area may become unlivable. Without the necessities required by humans to live or adapt to an environment, they must find a new place to live.

When a species is continually crowded out of its habitat by development or its habitat is divided and it can't reproduce, its numbers drop through *habitat fragmentation.* Adjacent habitats have overlapping boundaries called *ecotones* with higher species diversity and biological density than at a community's center. This increased diversity or *edge effect* allows some species to survive who couldn't live at the heart of the habitat. In fact, when ecotones are destroyed or changed, both edge and inner habit species are impacted.

When this happens, a species is said to be *endangered.*

Endangered species are those species threatened with extinction (e.g., Florida panther and California condor).

Sometimes habitat loss is so severe or happens so quickly, it results in a species being eliminated from the planet. This happened to the dinosaurs. Scientists are still trying to decide what caused the mass extinction and there are a lot of theories, but except for in Hollywood movies, huge dinosaurs no longer roam the Earth.

A species no longer living, anywhere on Earth, is said to be *extinct.*

Extinction takes place naturally, because for some species to succeed, others must fail. Since life began, about 99% of Earth's species have disappeared and, on several occasions, huge numbers have died out fairly quickly. The most recent of these mass extinctions, about 65 million years ago, swept away the dinosaurs and many other forms of life. Though not extinct as a result of human actions, the dinosaurs are a good example of a large number of species unable to adapt to environmental changes.

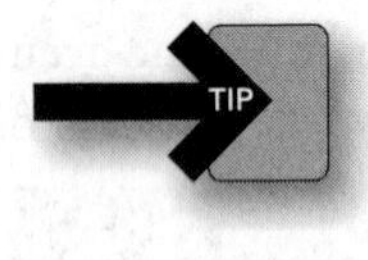

In the United States, conservation efforts were strengthened when the *Marine Mammals Protection Act* (1972), *Endangered Species Act* (1973), and *Convention on International Trade in Endangered Species of Wild Flora and Fauna* (CITES, 1973) were passed. These policies protected land and marine animals and made it a crime to hunt, capture, or sell species considered endangered or threatened.

Local extinction takes place when every member of a specific population in a specific area has died. Table 9.2 shows the number of species evaluated and those placed on the Endangered Species List by the World Conservation Union. For the past 50 years, the *World Conservation Union's Species Survival Commission* (SSC) has been ranking the conservation status of species, subspecies, varieties, and selected subpopulations worldwide, to pinpoint groups threatened with extinction. To promote their conservation efforts, the SSC has the most current, objective, scientifically based information on the status of globally threatened biodiversity available. The collected data on species rank and distribution gives policy makers solid information with which to make informed decisions on preserving biodiversity at all levels.

In 2014, the World Wildlife Fund's watch list included chimpanzees, dolphins, polar bears, tigers, gorillas, pandas, elephants, rhinos, and whales. A few species that have approached extinction or become completely extinct include the *Gorilla beringei beringei* (African mountain gorilla), *Pyrenean ibex* (European goat), *Canis rufus floridianus* (Florida wolf), and *Hippopotamus madagascariensis* (Madagascan hippo). Global extinction happens when every member of a species on the Earth has died. The passenger pigeon and the dodo are examples of globally extinct birds. Extinct is forever.

Wetlands

Wetlands provide the habitat for richly diverse populations. Once considered unimportant, wetlands are now known to support important and extensive ecosystems. Wetland plants convert sunlight into plant material or biomass and provide food to many different kinds

Table 9.2 Species are becoming endangered through habitat loss, pollution, and poaching.

	NUMBER OF SPECIES	NUMBER OF SPECIES ASSESSED (2014)	NUMBER OF THREATENED SPECIES (2014)
Mammals	5,513	5,513	1,199
Birds	10,425	10,425	1,373
Reptiles	9,952	4,256	902
Amphibians	7,286	6,410	1,961
Fishes	32,800	11,323	2,172
Insects	1,000,000	4,980	954
Molluscs	85,000	7,109	1,929
Crustaceans	47,000	3,164	725
Corals	2,175	856	235
Others	68,658	453	65
Mosses	16,236	102	76
Ferns	12,000	359	194
Lichens	17,000	2	2
Mushrooms	31,496	1	1
Total	1,345,541	54,953	11,788

of aquatic and land animals, supporting the aquatic food chain. Wetlands, often protected, also provide moisture and nutrients needed by plants and animals alike.

Wetlands are low, soggy places where land is constantly or seasonally soaked, or even partly underwater.

Wetlands, transitional areas between land and marine areas, can be swamps, bogs, peat lands, fens, marshes, or swamp forests. The water table is above, even with, or near the land's surface. Wetland soils hold large amounts of water and their plants are tolerant of occasional flooding.

About 60% of U.S. major commercial fisheries use estuaries and coastal marshes as nurseries or spawning sites. Migratory waterfowl and other birds also rely on wetlands for homes, stopovers, and food.

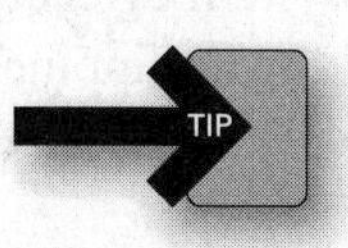

Wetlands are home to more than 600 animal species and 5,000 plant species. In the United States, nearly 50% of the species on the endangered animals list, and 25% of the plants, live in or rely upon wetlands. One-half of U.S. migratory birds are dependent on wetlands.

Internationally, wetlands are taking a hit as well. In Canada, which contains 25% of the world's wetlands, 15% of the wetlands have been lost. Germany and the Netherlands have lost over 50% of their wetlands.

Ocean Residents

Besides overfishing and introduction of alien marine species, there are environmental concerns for nearly every resident of the world's oceans. Everything from sharks, whales, and dolphins, to jellyfish, tube worms, and kelp beds has been impacted by ocean pollution. Since even the smallest members (microorganisms) of the food web are affected by chemicals, turbidity, and temperature increases, pollutants cause a domino effect as larger and larger species are impacted.

Hotspots

In 1988, British ecologist Norman Myers described the biodiversity hotspot idea. Although tropical rain forests have the highest extinction rates, they aren't the only places at risk. Myers pointed out a resource problem facing ecologists. They couldn't save everything at once and needed a way to identify areas with endangered species.

Globally, there are hundreds of species facing extinction because of habitat destruction and loss. Myers identified 18 high-priority areas where habitat cover had already been reduced to less than 10% of its original area or would be within 20 to 30 years. These regions make up only 0.5% of the Earth's land surface, but provide habitat for 20% of the world's plant species facing extinction.

TIP

Two factors weigh heavily in identifying a hotspot: (1) high diversity of endemic species and (2) significant habitat impact and alteration from human activities.

Plant diversity is the biological basis for hotspot designation. To qualify as a hotspot, a region must support 1,500 endemic plant species, of the total worldwide population. Existing natural vegetation is used to assess human impact in a region.

> An ecological region that has lost more than 70% of its original habitat is known as an environmental *hotspot.*

Since plants provide food and shelter for other species, they are used in rating an area as a hotspot. Commonly, the diversity of endemic birds, reptiles, and animals in hotspot areas is also extremely high. Hotspot animal species are found only within the boundaries of the hotspot, since they are often specifically adapted to endemic plant species as their main food source.

In hotspot designations by world conservation agencies, 25 biodiversity hotspots, containing 44% of all plant species in roughly 1.0% of the planet's land area, were listed. Hotspots target regions where the extinction threat is the greatest to the highest number of species. This allows biologists to focus cost-effective efforts on critical species.

Endemic species have been isolated over geological time. Islands, surrounded by water, have the most endemic species. In fact, many of the world's hotspots are islands. Topographically different areas like mountain ranges allow the greatest ecosystem diversity.

Several hotspots are tropical island archipelagos, like the Caribbean and the Philippines, or big islands, like New Caledonia. However, other hotspots are continental islands isolated by surrounding deserts, mountain ranges, and seas.

Peninsulas are key regions for hotspots. They are similar to islands and some, like Mesoamerica, Indo-Burma, and the Western Ghats in India, were islands at some time in the past. Other hotspots are landlocked islands isolated between high mountains and the sea. The Andes Mountains, which separate South America from north to south, are an impassible barrier to many species. On the western coast, the lowlands support a thin ecosystem, isolated from the eastern side of the continent.

The Cape Floristic Province in South Africa is isolated by the extreme dryness of the Kalahari, Karoo, and Namib deserts, and large rivers like the Zambezi and the Limpopo.

Why Are Hotspots Fragile?

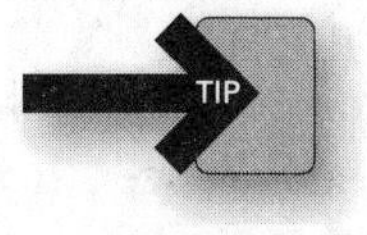

Island ecosystems are particularly fragile because they are rarely exposed to outside influences. Ecologists have found that most extinct species were island species and not widely spread. They lived in specific isolated habitats. Once a one-of-a-kind population is gone, the species is lost forever.

Isolated species lose their defenses over time, because they are only exposed to a limited number of other species. When they have to compete with new, previously unknown species, they can't adapt fast enough. This is especially true if the new species is highly competitive and adaptable.

For example, large extinct birds like the *moa* and *dodo*, which had no predators on the remote Australian continent, lost their ability to fly. When humans and other predators arrived, they were easy targets and quickly dropped in numbers.

Since many global hotspots are beautiful and unique, humans have been drawn to their natural diversity throughout human history. Ecosystems and landscapes were changed, first by hunter-gatherers, then by farmers and herdsmen, and most extensively by the global growth and sale of agricultural crops. During the past 500 years, many species have been hunted to the last individual.

Currently, growing human populations in world hotspots add to species' decline by the introduction of nonnative species, illegal trade in endangered species, industrial logging, slash and burn agricultural practices, mining, and the construction of highways, dams, and oil wells. Eleven hotspots have lost at least 90% of their original natural vegetation, and three of these have lost 95%.

Today, the world's regions considered the "hottest of the hot spots" are Madagascar and the Indian Ocean islands, the Philippines, Sundaland, the Atlantic Forest, and the Caribbean. These five hotspots have the most unique biodiversity and are at extreme risk of losing it without immediate and effective conservation.

Conservation

Since hotspots have the highest concentrations of unique biodiversity on the planet, they are also at the greatest risk. We must preserve hotspot species to prevent a domino effect of ecosystem extinction. Knowledge and tools to protect hotspots must be in place, as well as ongoing updates of political, social, and biological conditions associated with hotspots.

Information on hotspot species is being collected, and biological evaluations made in little-understood land, freshwater, and ocean ecosystems. Teams of international and regional biologists are performing hotspot assessments. Field station networks of all the world's main tropical areas are being set up to monitor biodiversity.

Solutions

Different conservation methods are important to protect hotspot biodiversity. These vary from the creation of protected areas to alternatives like ecotourism. Educating people at the local and national levels is also important. Governmental policies and awareness programs, with improved business practices to protect against ongoing biodiversity loss, are critical.

Strengthening existing conservation efforts lessens potential climate destabilization and offers greater resilience against weather disasters that threaten both people and habitat.

Creating protected areas and conservation regions, and improving the administration of over 55 million acres of parks and protected areas in hotspots and wilderness areas are crucial to ensuring continued biodiversity.

Species' habitat ranges adjust to climate change, which impacts ecologists' ability to protect them in existing parks. Range boundary shifts due to temperature increases have been taking place for over 75 years. To lower extinction risks connected with global warming, conservation methods must be developed to address this problem.

Many medicines come from plants and fungi. New species in today's hotspots may hold the key to research and treatments for human disorders like emphysema and cancer. It is important to protect these valuable resources. However, conserving biodiversity in hotspots worldwide is not an easy job. No one country or organization can do it alone. Everyone has to work together. As the world's population continues to climb, environmental issues will become critical for more and more species, including our own.

Remote Sensing

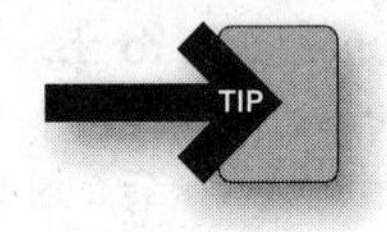

When environmentalists make observations, take measurements directly, or collect samples, it's called *in situ* data collection or *field sampling.* Field sample collection and analysis are done at a sample site, but when too many people take samples from the same place, the area can get stomped down and disturbed. This is why national parks and forests ask people to stay on trails. They want to protect pristine areas.

When scientists study an area and want to avoid disturbing the environment, they use remote sensing instruments aboard aircraft, high-altitude balloons, and satellites. Much of this technology was developed as stealth imaging during war time or for space exploration by NASA. The environmental benefits from this cutting-edge technology make it possible to accurately image an object or environment to within a meter.

The measurement or study of an object, area, or event by a distant recording device is known as *remote sensing.*

The *Advanced Spaceborne Thermal Emission and Reflection Radiometer* (ASTER) is a remote sensing instrument. It is located aboard *Terra,* a satellite launched in 1999 as part of NASA's *Earth Observing System* (EOS). ASTER is a joint project between NASA; the Japanese Ministry of Economy, Trade, and Industry; and the Earth Remote Sensing Data Analysis Center. ASTER data are used to draw detailed maps of land surface temperature, reflectance, and elevation.

Different remote sensing instruments can gather information about temperature, chemistry, photosynthetic ability, moisture content, and location. Table 9.3 shows some of the geographical and ecological characteristics that can be observed with remote sensing.

Habitat hotspots can also be observed with satellite imagery and aerial photography. For example, moisture information helps track changes in vegetation over time. Sensors record electromagnetic energy from vegetation without having to collect samples. When analysis is needed, scientists use remote sensing to direct them to problem areas so that samples of affected plants and trees can be taken.

Maps can be made of images taken at different times. These maps help scientists and policy makers by showing deforestation and habitat loss, as well as other biodiversity threats, such as forest fires, illegal logging, and construction development.

Table 9.3 Important land and water characteristics can be determined by remote sensing.

BIOLOGICAL AND PHYSICAL CHARACTERISTICS	REMOTE SENSING SYSTEMS
x, *y* Location	Aerial photography, Landsat, SPOT HRV, Space Imaging IKONOS, ASTER, Radarsat, ERS-1,2 microwave, ATLAS
z Topographic/depth measurement	Aerial photography, TM, SPOT, IKONOS, ASTER, Radarsat, LIDAR systems, ETM
Vegetation (*chlorophyll concentrations, biomass, water content, absorbed photosynthetic radiation, phytoplankton*)	Aerial photography, TM, SPOT, IKONOS, ETM, Radarsat, TM Mid-IR, SeaWiFS, AVHRR, IRS-1CD
Surface temperature	GOES, SeaWiFS, AVHRR, TM, Daedalus, ATLAS, ETM, ASTER
Soil moisture	ALMAZ, TM, ERS-1,2 Radarsat, Intemap Star 3i, IKONOS, ASTER
Evapotranspiration	AVHRR, TM, SPOT, CASI, ETM, MODIS, ASTER
Atmosphere (*chemistry, temperature, water vapor, wind speed/direction, energy input, precipitation, clouds, and particulates*)	GOES, UARS, ATREM, MODIS, MISR, CERES
Reflectance	MODIS, MISR, CERES
Ocean (*color, biochemistry, phytoplankton, depth*)	POPEX/POSEIDON, Sea WiFS, ETM, IKONOS, MODIS, MISR, ASTER, CERES
Snow and sea ice (*distribution and characteristics*)	Aerial photography, AVHRR, TM, SPOT, Radarsat, SeaWiFS, ICONOS, ETM, MODIS, ASTER
Volcanoes (*temperature, gases, eruption characteristics*)	ATLAS, MODIS, MISR, ASTER
Land use	Aerial photography, AVHRR, TM, SPOT, IRS-ICD, Radarsat, Star 3i, IKONOS, MODIS

*In part from Jensen, John R., 2007, *Remote Sensing of the Environment: An Earth Resource Perspective*, 2nd Ed., Upper Saddle River, NJ: Prentice Hall.

› Review Questions

Multiple-Choice Questions

1. A complex community of plants, animals, and microorganisms linked by interacting energy and nutrient flows is called a(n)

(A) atmosphere
(B) ecosystem
(C) niche
(D) suburb
(E) species

2. The world's oceans make up what percent of the planet's biosphere?

(A) 27%
(B) 44%
(C) 62%
(D) 75%
(E) 99%

3. Which diverse ecosystem absorbs high flow and releases water slowly?

(A) High plains
(B) Wetlands
(C) Arctic
(D) Rocky Mountains
(E) Old-growth forest

4. Madagascar and the Indian Ocean islands, the Philippines, Sundaland, the Atlantic Forest, and the Caribbean are all considered

(A) hotspots
(B) expensive vacation spots
(C) sustainable use areas
(D) arid regions
(E) mountainous

5. Aircraft, high-altitude balloons, and satellites are all used in

(A) birthday parties
(B) field testing
(C) remote sensing
(D) acrobatic air shows
(E) high school science labs

6. The vertical range that contains the biosphere is roughly

(A) 1,000 meters high
(B) 5,000 meters high
(C) 10,000 meters high
(D) 20,000 meters high
(E) 40,000 meters high

7. When a species like the dodo bird becomes extinct, it is said to be

(A) in remission
(B) hibernating for the winter
(C) not important to the ecosystem
(D) gone for 10 years and then returns
(E) gone forever

8. The total area in which a plant, animal, insect, or other organism travels in its lifetime determines its

(A) life span
(B) range
(C) personality type
(D) itinerary
(E) habitat

9. ASTER information is used to draw detailed maps of land surface temperature, elevation, and

(A) salt flats
(B) mountain algae populations
(C) reflectance
(D) honeybee populations
(E) shipping lanes

10. Adjacent habitats have overlapping boundaries called

(A) wetlands
(B) topography
(C) ecological succession
(D) ecotones
(E) secondary succession

11. Over time, a previously untouched area like a volcanic flow will be populated by different species creating soil, cover, shade, and food resulting in a(n)

(A) primary succession of development
(B) secondary succession of development
(C) tertiary succession of development
(D) overcrowded ecosystem
(E) ecotone

12. A hotspot is an ecological region that has lost

(A) 10% of its original habitat
(B) 20% of its original habitat
(C) 45% of its original habitat
(D) 60% of its original habitat
(E) over 70% of its original habitat

13. The most important factor currently affecting many species is

(A) loss of habitat
(B) predation
(C) food supply
(D) climate
(E) disease

14. The use of resources in a manner which protects a species or environment without causing long-term loss is known as

(A) evolutionary use
(B) clear cutting
(C) sustainable use
(D) extinction
(E) speciation

15. When no two species can fill the same niche at the same time and the weaker species must fill a smaller niche, relocate, or die, it is known as

(A) survival of the fittest
(B) Gause's principle
(C) the conservation of species
(D) a keystone species
(E) climax succession

16. When a species' numbers drop continuously because of habitat destruction, the species is said to be

(A) extinct
(B) poor predators
(C) protected
(D) not evolutionarily hearty
(E) endangered

17. Mutualism is a type of symbiosis where

(A) neither species benefits
(B) both species benefit
(C) one species benefits and one is unaffected
(D) one species is harmed and one benefits
(E) both species are harmed

18. When a species has no competitive limitations, it has a

(A) realized niche
(B) developed niche
(C) fundamental niche
(D) successive niche
(E) population overshoot

19. When an ecosystem reaches its final stage of balanced species development, it is called a(n)

(A) climax community
(B) population
(C) pioneer species
(D) initial niche
(E) extinct zone

› Answers and Explanations

1. B—Several different niches combine into an ecosystem.

2. E

3. B—Wetlands support important and extensive ecosystems by providing food.

4. A—These regions have lost more than 70% of their original habitat.

5. C—Different remote sensing instruments (e.g., infrared sensors) allow different data gathering.

6. D

7. E—Every last member of the species is dead and none remain to reproduce.

8. B—A species' geographical range can extend across wide areas, depending on conditions.

9. C—These data change depending on the amount of trees, cloud cover, rainfall, and so forth.

10. D—Higher species diversity and density contribute to an edge effect.

11. A—This is done by pioneer species.

12. E

13. A—Habitat loss often affects the other factors causing a species' decline.

14. C—Impacting an area to a recoverable extent.

15. B—The species settling for a smaller niche is said to have a realized niche.

16. E

17. B—They get a positive benefit from their alliance and would compete less well without it.

18. C

19. A—It has progressed through the various growth stages and is mature.

Free-Response Question

1. Scientists have observed severe declines in honeybee populations. They call this colony collapse disorder. Without honeybees, flowers are not pollinated, crops don't thrive, and overall plant growth suffers. In fact, apple growers are almost totally dependent on insects for pollination, with honeybees responsible for over 90%. Pumpkin growers need bees to pollinate their plants too. On average, growers paid \$105–\$140 per colony in 2014, compared to \$55 and \$65 per colony in 2008 due to the decline in available bees. In fact, nearly half of all honey bee colonies have vanished due to pesticides, mites, and disease.

(a) From the preceding information, can honeybees be considered a keystone species in certain agricultural ecosystems?
(b) Describe a possible effect of drastic honeybee population declines as a species.
(c) How could the population drop in honeybee population affect the habitats of other local denizens?
(d) Ultimately, where does all food come from (i.e., what major components are necessary?)?

Free-Response Answers and Explanations

1.

a. Because their presence is vital to the health of other plant and animal species, bees can be considered a keystone species. Without honeybees both crops and flowers can fail leading to drastic changes in the surrounding environment. This affects all creatures whose livelihood depends on the existence of the crops and flowers that honeybees make possible. The overall suffering of plant growth affects not only the flora but the fauna that nest in, hide in, and feed on plants to survive.

b. Genetic diversity is important for the overall health of a species. Population reduction reduces this diversity and can cause a bottleneck effect, thus affecting the prospects for the species to adapt to different environmental pressures.
c. Because the overall health of plant life in ecosystems reliant on honeybees suffers, so do the habitats that such plant life provides for other animals. Loss of habitat dramatically affects the health of any given species because they then cannot find proper nutrients, water, sunlight, shelter, and living space.
d. Solar energy, water, soil, photosyntheis, and oxygen.

❯ Rapid Review

- Loss of habitat is perhaps the most important factor that affects species.
- Wetlands are the transitional areas between land and marine areas.
- Land-based ecosystems, known as biomes, are classified by rainfall and climate.
- An ecosystem is a complex community of plants, animals, and microorganisms linked by energy and nutrient flows that interact together and with their environment.
- An animal or plant with a specific relationship to its habitat or other species, filled by it alone, occupies an ecological niche.
- Sustainable use is the use of resources in a way that protects the numbers and complexity of a species or environment without causing long-term loss.
- The biosphere includes the hydrosphere, crust, and atmosphere. It is located above the deeper layers of the Earth.
- The total genetic makeup of a population is called its gene pool.
- Species reach levels of specialization through adaptation (i.e., changing as the environment changes) and evolution (i.e., genetic changes in species over time).
- Random DNA protein breaks or other changes are called genetic drift.
- When the frequency of certain traits is narrowed from the original population, it causes a bottleneck effect.
- When only a fraction of original genetic material and its diversity is represented in a new group, it is known as a founder effect.
- The Hardy-Weinberg principle states that in a stable population, the frequency of genotypes and alleles (parts of genetic material) will remain constant.
- A species is a group of intrabreeding organisms unable to interbreed with a different species.
- Allopatric speciation often results from reproductive isolation in animal and plant species.
- Sympatric speciation, or the evolution of two different species from a single species, is more common in plants.
- When two species compete and the stronger, better adapted species wins, the process is called competitive exclusion.
- When a species has no competitive limitations, it has its fundamental niche.
- When a species must settle for a smaller niche than it normally would have because of competition, the niche is called a realized niche.
- Gause's principle explains that no two species can fill the same niche at the same time, and the weaker species will fill a smaller niche, relocate, or die off.
- When an ecosystem reaches its final stage of balanced species development, it is called a climax community.
- Predation occurs when one species serves as food for another species.

- Symbiosis is the close, extended relationship between organisms of different species that may (↑) or may not (↓) benefit each participant. There are three types of symbiotic relationships:
  - Mutualism. Both species benefit (↑↑).
  - Commensalism. One species benefits (↑) and one species is fairly unaffected (~).
  - Parasitism. One species benefits (↑) and one species is harmed (↓).
- A related mechanism of passive competition is camouflage and mimicry.
- Endemic species, naturally occurring in only one area or region, are unique to that specific region.
- The total area in which a plant, animal, insect, or other organism may travel in its lifetime is considered its range.
- The first colonizers to a site (e.g., moss or lichens) are called pioneer species.
- A species around which an entire ecosystem is dependent is known as a keystone or foundational species.
- Endangered species are those species threatened with extinction (like the Florida panther and California condor).
- A species that is no longer living, anywhere on Earth, is said to be extinct.
- The area in which an animal, plant, or microorganism lives and finds nutrients, water, sunlight, shelter, living space, and other essentials is called its habitat.
- Wetlands are low, soggy places where land is constantly or seasonally soaked, or even partly underwater.
- An ecological region that has lost more than 70% of its original habitat is known as an environmental hotspot.

Natural Cycles and Energy Flow

IN THIS CHAPTER

Summary: Although the Earth appears to consist of one big environment, it is composed of many smaller, interconnected systems or cycles. The carbon, calcium, hydrologic, and other cycles allow other systems to function. Plants and animals get food through interactions with these cycles in a food web.

Keywords

✪ Matter, carbon, organic, conservation of matter, calcium, phosphorus, sulfur, hydrologic cycle, evapotranspiration, photosynthesis, residence time, biomineralization, food web, trophic level, biomass

Geochemical Cycle

A geochemical cycle is like an element's life cycle. As it moves from one place to another, it takes on different forms. In this chapter, we will take a look at several geochemical cycles including the calcium, carbon, and hydrologic cycles. These major Earth cycles have intricate and complex interrelationships that exist at all levels of diverse ecosystems.

Conservation of Matter

All things that take up space and have mass are known as *matter*. Matter exists in three forms, solid, liquid, and gas. Except in unusual circumstances, matter cannot be created or destroyed. It is recycled and transformed in many different combinations, systems, and organisms over and over again. It doesn't disappear; it just moves into another form and cycle.

Conservation of matter includes the idea that matter is neither created, nor destroyed, but recycled through natural cycles.

Residence Time

To understand natural cycles, we need to understand *residence time.* Residence time equals the average amount of time a chemical element (e.g., carbon, calcium, or phosphorus) spends in a geological reservoir or cycle.

Carbon dioxide, at about 5% of the atmosphere, has a residence time of 10 years. In contrast, oxygen, which makes up around 20% of the atmospheric volume, has a residence time of 6,000 years. Sulfur dioxide, a very minor atmospheric player, has a residence time of hours to weeks. Amazingly, nitrogen, in amounts that equal 75% of the total atmospheric gases has a residence time of 400 million years. Dinosaurs probably breathed some of the same nitrogen molecules we are breathing today!

Calcium spends its residence time in the atmosphere, oceans, crust, and mantle. Carbon spends its residence time in the atmosphere; oceans; sedimentary, igneous, and metamorphic rocks; and the biosphere.

Carbon

Carbon is the fourth most abundant element in the universe, after hydrogen, helium, and oxygen. Known as the building block of life, carbon is the foundational element of all *organic* substances, from graphite to fossil fuels to DNA. On the Earth, carbon cycles through the land, biosphere, ocean, atmosphere, and the Earth's interior in a major biogeochemical cycle.

Organic matter is made up of carbon-containing material from living or nonliving material and includes the organic parts of soil.

The carbon cycle has many different storage spots, also known as reservoirs or *sinks,* where carbon exchanges take place. Carbon can be stored in the atmosphere, oceans, and soil as carbon dioxide, oil, coal, or biomass. The carbon cycle, shown in Figure 10.1, is divided into two types, the *geological carbon cycle,* which has been going on for millions of years, and the *biological carbon cycle,* which stretches from days to thousands of years.

Geologists believe the total amount of carbon that cycles through today's Earth systems was around at the formation of the solar system.

Geological Carbon Cycle

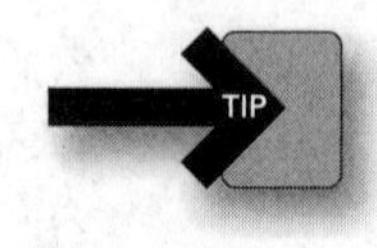

In the geological carbon cycle, carbon moves between rocks and minerals, seawater, and the atmosphere through weathering. Carbon dioxide in the atmosphere reacts with water and minerals to form calcium carbonate. Calcium carbonate rock (limestone) is dissolved by rainwater through erosion and then carried to the oceans. There, it settles out of the ocean water, forming sedimentary layers on the sea floor. Then, through plate tectonics, these sediments are subducted underneath the continents. With the extreme heat and pressure deep beneath the Earth's surface, the limestone melts and reacts with other minerals, freeing carbon dioxide. This carbon returns to the atmosphere as carbon dioxide during volcanic eruptions, completing the carbon cycle.

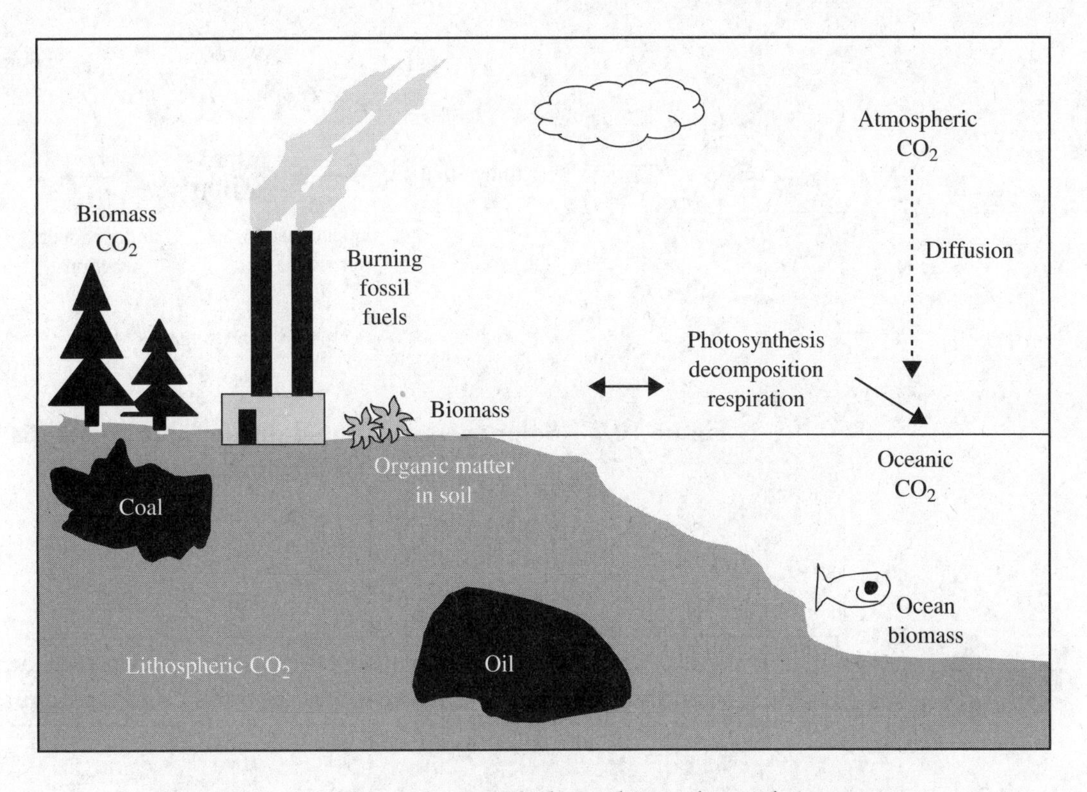

Figure 10.1 The carbon cycle has a big industrial component.

The balance between weathering, subduction, and volcanism controls atmospheric carbon dioxide concentrations over geological time. Some geologists have found that the oldest geological sediments point to atmospheric carbon dioxide concentrations over 100 times current levels.

Conversely, ice core samples from Antarctica and Greenland make glaciologists think that carbon dioxide concentrations during the last ice age were only about one-half of today's levels.

The amount of carbon stored and exchanged at each step in the cycle controls whether a certain sink is increasing or decreasing. For example, if the ocean absorbs 2 gigatons more carbon from the atmosphere than it releases in any one year, then atmospheric storage will decrease by the difference. Additionally, the atmosphere interacts with plants, soils, and fossil fuels. Everything is intimately interconnected.

The carbon cycle is a closed system. All carbon is squirreled away on the planet somewhere. Geologists are trying to balance out the global carbon equation. When all the sinks are estimated and added up, both sides of the equation should be equal. As population increases and global resources are challenged, experiments in this area are going to be more and more important.

Biological Carbon Cycle

The biosphere and all living organisms play a big role in carbon movement into and out of the land and ocean through *photosynthesis* and *respiration* processes. Photosynthesis defines the series of reactions in plants, bacteria, and algae that capture visible light wavelengths (0.4 to 0.7 μm) and transform light into chemical energy needed for organic molecules to bond. Figure 10.2 illustrates how photosynthesis works.

On this planet, nearly every living thing depends on the creation of sugars and carbohydrates from photosynthesis and the metabolism (respiration) of those sugars to support biological growth and reproduction.

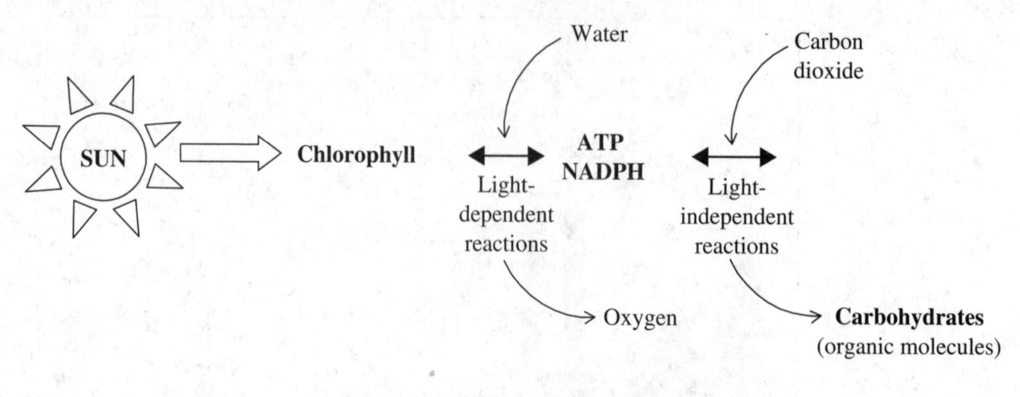

Figure 10.2 Solar energy captured during photosynthesis is used by nearly all life on Earth.

The *biological carbon cycle* occurs when plants absorb carbon dioxide and sunlight to make glucose and other sugars (carbohydrates) to build cellular structures.

Plants and animals use carbohydrates during respiration, the opposite of photosynthesis. Respiration converts biological (metabolic) energy back to carbon dioxide. As a process pair, respiration and decomposition (respiration by bacteria and fungi) restore biologically fixed carbon back to the atmosphere. Yearly carbon levels taken up by photosynthesis and sent back to the atmosphere by respiration are 1,000 times higher than carbon transported through the geological cycle each year.

We've seen how photosynthesis and respiration play a big part in the long-term geological carbon cycling. Land plants pull carbon dioxide from the atmosphere. In the oceans, the calcium carbonate shells of dead phytoplankton sink to the sea bed and form sediments. When photosynthesis is higher than respiration, organic matter gradually builds over millions of years and forms coal and oil deposits. These biologically regulated activities characterize atmospheric carbon dioxide removal and carbon storage in geological sediments.

Balance

Carbon is stored in the following major storage reservoirs:

- Organic molecules in living and dead organisms found in the biosphere
- Atmospheric carbon dioxide
- Organic matter in soils
- Fossil fuels in the lithosphere, sedimentary rock (limestone, dolomite, and chalk)
- Dissolved atmospheric carbon dioxide in the oceans
- In the calcium carbonate of marine creatures' shells

Over geological history, the amount of carbon dioxide found in the atmosphere has dropped. It is hypothesized that when the Earth's temperatures were a bit higher, millions of years ago, plant life was plentiful because of the greater concentrations of atmospheric carbon dioxide. As time went on, biological mechanisms slowly locked some of the atmospheric carbon dioxide into fossil fuels and sedimentary rock. This carbon balancing process has kept the Earth's average global temperature from huge swings over time.

Table 10.1 Carbon is stored in various areas above and below the earth.

CARBON STORAGE	QUANTITY (BILLIONS OF METRIC TONS)
Atmosphere	580 (1,700)–800 (2,000)
Organic (soil)	1,500–1,600
Ocean	38,000–40,000
Ocean sediments and sedimentary rocks	66,000,000–100,000,000
Land plants	540–610
Fossil fuels	4,000

Geologists are interested in carbon because it is such a versatile element. Not only does carbon exist in the air, land, and sea but humans are made up of approximately 50% carbon by dry weight. Environmental chemists study different ecosystems with carbon balancing accounts using crop productivity, food chains, and nutrient cycling measurements.

The *carbon cycle* involves the Earth's atmosphere, fossil fuels, oceans, soil, plants, and animal life of terrestrial ecosystems.

In addition, carbon dioxide is the main atmospheric greenhouse gas thought to be a result of human activities. Until alternative power sources, like solar power, are developed and used more, atmospheric carbon dioxide increases will result mostly from burning fossil fuels.

Geologists look for patterns when trying to understand seasonal carbon drops and gains in atmospheric carbon dioxide. We've seen how global photosynthesis and respiration have to balance or carbon will either accumulate on land or be released to the atmosphere. Measuring year-to-year changes in carbon storage is tough. Some years have more volcanic eruptions with extra carbon in the air, while other years or decades have less.

However, some measurements are straightforward. The clearing of forests for crops, for example, is well documented, both historically and from satellite data. When forests get a chance to grow back on cleared land, they pull carbon from the atmosphere and start saving it up again in trees and soils. The change between total carbon released to the atmosphere and the total pulled back down governs whether the land is a supplier or reservoir of atmospheric carbon.

Atmospheric carbon = fossil fuels + land use changes − ocean uptake − unknown carbon deposit.

When considering the global carbon equation between the atmosphere, fossil fuels, and the oceans, the global carbon tally is not completely known. Research is ongoing to discover the location of unknown carbon reservoirs.

Calcium

Calcium makes up roughly 3.4% of the Earth's crust and has been around since the formation of the Earth. It is found in igneous rocks as *calcium silicates,* and in sedimentary and metamorphic rocks as *calcium carbonates.* When rock weathering takes place by acid rain or plant growth and decay, calcium interacts with water and is transported to another location.

Water helps calcium move from the land to the oceans. High concentrations of dissolved calcium and/or magnesium in fresh water cause *hard water.* When these minerals are concentrated in water, around 89 to 100 parts per million, they don't react well with soap. In fact, mineral rings form in bathtubs, and laundered clothes take on a gray color from undissolved soap scum. Undissolved minerals in hard water are also deposited in plumbing, coffee pots, and steam irons. Frequently, people living in hard water areas use water softeners—chemicals that remove calcium and magnesium ions in an exchange with sodium ions.

Hard water can have benefits. In the aquatic environment, calcium and magnesium help keep fish from absorbing metals like lead, arsenic, and cadmium into the bloodstream through their gills. Therefore, the harder the water, the less potential for toxic metals to be absorbed by fish. In seawater, calcium concentrations are 100 to 1,000 times higher than land levels, and even greater concentrations are found in deeper, colder waters with little circulation. Calcium can reside roughly a million years in the ocean before it appears on land again. Calcium ions stay in ocean water until they are precipitated out as calcium carbonate.

Although upper ocean levels are highly saturated with calcium and carbonate ions, saturation depends on location and conditions. Photosynthesis and temperature affect warm, shallow water, lowering levels of carbon dioxide. These conditions allow calcium carbonate to precipitate either inorganically or through aquatic organisms.

When calcium carbonate is used by marine inhabitants to build shells, it is called *biomineralization.* As these organisms die, their hard shells sift down to the ocean's floor and gather or dissolve depending on depth, temperature, and pressure. Shells sinking to the deepest parts of the ocean often redissolve because of higher carbon dioxide levels in the colder, deeper waters.

The dividing line separating an area where calcium carbonate dissolves and accumulates is called the *lysocline.*

The calcium carbonate deposited by microorganisms is often mixed with other ocean sediments or washed from the land depending on location. Birds, animals, and humans eat seafood and shellfish, discarding the shells. This returns calcium back to the Earth fairly rapidly.

However, most calcium is transported by plate tectonics. Crustal plates and continental land masses with their mountain-building movement help calcium carbonate deposits move toward the surface in the form of limestone or marble (if changed by pressure and temperature). Figure 10.3 illustrates the different compounds of the calcium cycle.

Soil

Calcium is taken up by plant roots either directly from the soil or from groundwater. When calcium is extracted from the soil via membrane permeability, both active and passive

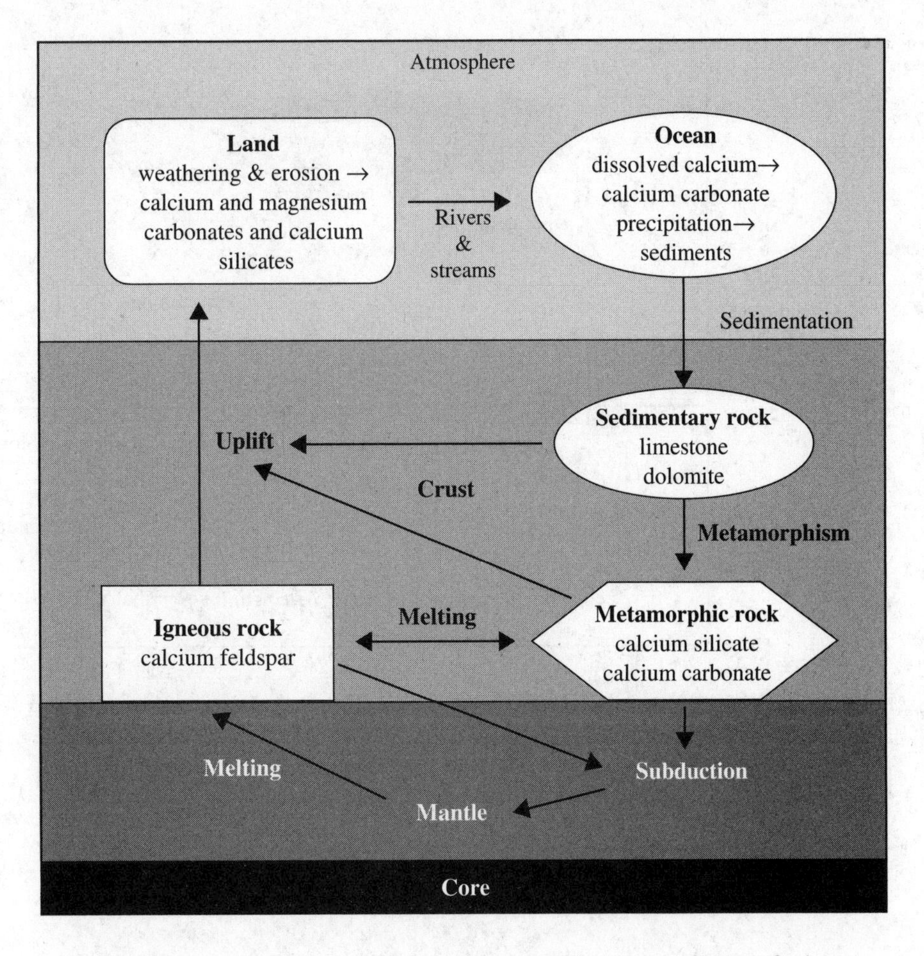

Figure 10.3 Calcium enters and exits sinks in a variety of places.

transport of ions takes place. Calcium is then transferred and stored in the leaves for a time. Calcium returns to the soil when leaves fall. It is stored in woody plant parts until it decays, is burned, or is consumed by an animal.

Nitrogen

Nitrogen (N_2) makes up 79% of the atmosphere. All life on Earth requires nitrogen-containing compounds (e.g., proteins) to survive. However, they can't easily use nitrogen in its gaseous form. To be used by a living organism, nitrogen must be combined with hydrogen and oxygen. Nitrogen is pulled from the atmosphere by lightning or nitrogen-fixing bacteria. During storms, large amounts of nitrogen are oxidized by lightning and mixed with water (rain). This falls and is converted into nitrates. Plants take up nitrates and form proteins.

Plants are consumed by herbivores or carnivores. When these consumers die (organic matter), nitrogen compounds are broken down into ammonia. Ammonia can be taken up by plants again, dissolved by water, or remain in the soil to be converted to nitrates (*nitrification*). Nitrates stored in soil can end up in rivers and lakes through runoff. It can also be

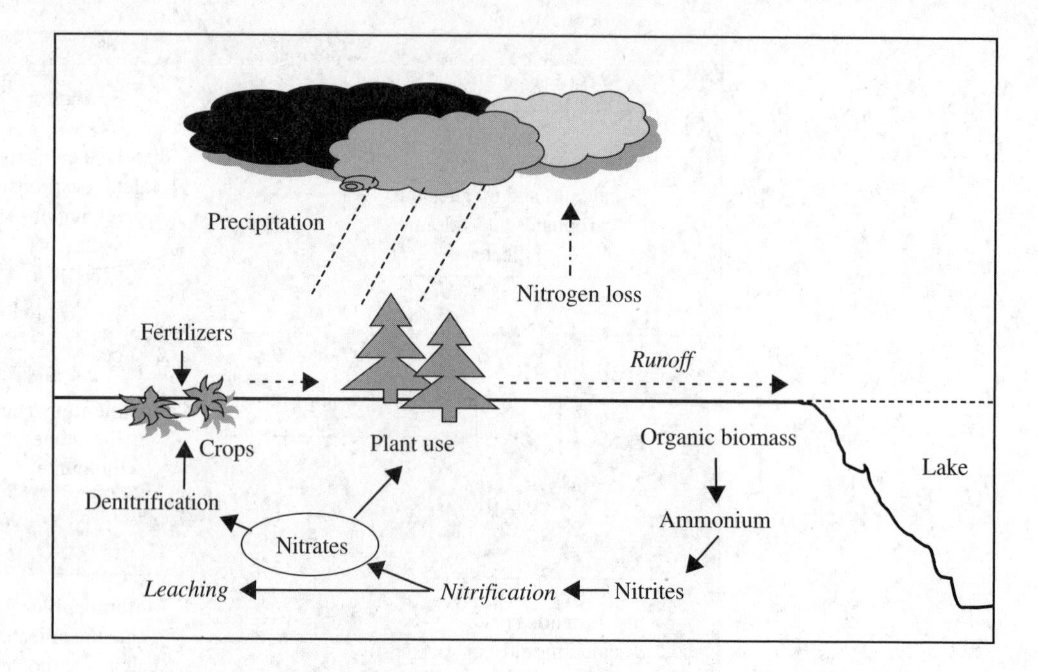

Figure 10.4 The nitrogen cycle is essential to all living systems.

changed into free nitrogen and returned to the atmosphere. Figure 10.4 gives you an idea of the nitrogen cycle.

Food Chains, Webs, and Trophic Levels

Photosynthesis is the foundation of all ecosystems. Plants are known as *primary producers* of biological material or *biomass.* Some organisms that produce biological material and eat plants are called *secondary producers.*

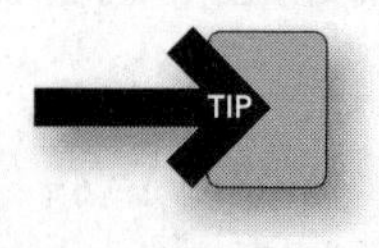

A *food chain* is the path that food follows. For example, if you eat a slice of cheese, you can trace it back to a cow (cheese is from milk) that ate plants for the energy and nutrients to make milk. Predatory species may eat several types of prey (food). A wolf may eat a rabbit that has just eaten a carrot. Then a few hours later, the wolf may come upon an injured deer which it kills and eats. The deer may have been grazing on grass before it was felled by the wolf. When individual food chains become interconnected, they are known as a *food web.* Adding microorganisms, worms, and insects, the web becomes even more complex.

A *trophic level* describes an organism's placement within a food chain or web. A carrot is at the producer level, while rabbits and chipmunks are primary consumers. There are several levels of consumers.

Some consumers, called *scavengers* (e.g., vultures and hyenas) feed from dead carcasses. *Detritivores* (e.g., ants and beetles) consume litter, debris (detritus), and dung, while *decomposers* (e.g., fungi and bacteria) finish off the process by breaking down and recycling organic matter. Figure 10.5 illustrates how organisms in an ecosystem get food (e.g., producer, primary, secondary, or tertiary consumers), as well as multilevel consumers (e.g., scavengers, parasites, and decomposers). The combination of all producers and consumers is called an *ecological pyramid.*

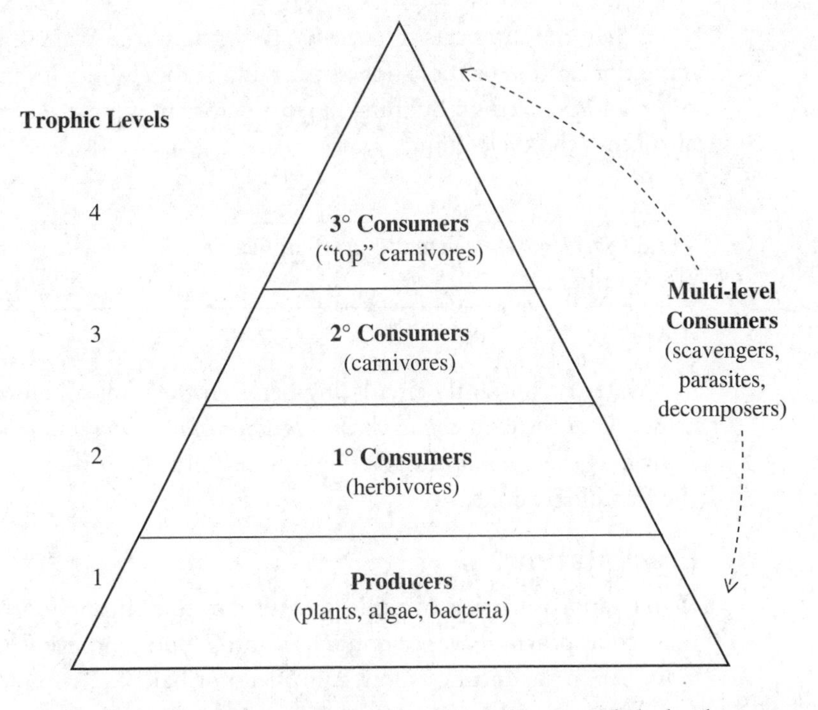

Figure 10.5 Lower organisms are producers, while herbivores and carnivores are consumers.

Mineral Cycles: Phosphorus and Sulfur

Phosphorus and sulfur are key minerals necessary for growth. At the cellular level, phosphorus is important in energy-transfer reactions. Too much phosphorus overstimulates plants and algae and is a major problem in water runoff and pollution.

The phosphorus cycle starts when the mineral leaches from rocks over a long time period. This form of phosphorus is taken in by producers, combined in organic molecules, and ultimately ingested by consumers. It goes back into the environment through decomposition of organic matter. Some phosphorus washes into rivers and eventually makes its way to the sea. Deep ocean sediments hold a significant amount of phosphorus.

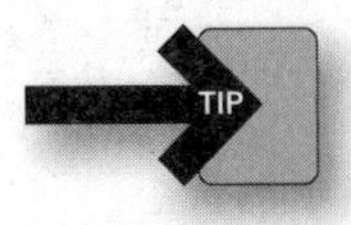

Sulfur plays an important part in proteins and controls the acidity of rainfall, surface water, and soil. Geological inorganic sulfur is found primarily in rock and such minerals as iron disulfate (iron pyrite) or calcium sulfate (gypsum). It gets into the air and water by weathering, gases from seafloor vents, and volcanism.

The sulfur cycle is affected by sulfur's many oxidation states, such as hydrogen sulfide (H_2S), sulfur dioxide (SO_2), sulfate ion (SO_4^{-2}), and elemental sulfur. Besides geological cycling, sulfur bacteria can anchor sulfur or release it into the environment. These bacteria are affected by pH, light, temperature, and oxygen concentrations.

The burning of fossil fuels also releases sulfur into the environment. Sulfur dioxide and other sulfur-containing atmospheric gases cause health problems, damage buildings and plants, and reduce atmospheric visibility.

Hydrologic Cycle

The hydrosphere, crust, and atmosphere combine to make up the biosphere. The hydrosphere includes all the water in the atmosphere and on the Earth's surface.

When the sun heats the oceans, the cycle starts. Water evaporates and then falls as precipitation in the form of snow, hail, rain, or fog. While it's falling, some of the water evaporates or is sucked up by thirsty plants before the rest soaks into the ground. The sun's heat also keeps the cycle going.

The *hydrologic cycle* is made up of all water movement and storage throughout the Earth's hydrosphere.

Water is constantly circulating between the atmosphere and the Earth and back to the atmosphere through *condensation, precipitation, evaporation,* and *transpiration.* This cycle is known as the *hydrologic cycle.* Figure 10.6 illustrates the many ways water moves through the hydrologic cycle.

Precipitation

Water vapor is carried by wind and air currents throughout the atmosphere. When an air mass cools down, its vapor condenses into clouds and eventually falls to the ground as *precipitation* in the form of snow, rain, sleet, or hail.

Water can take a variety of paths and time periods to get back into the atmosphere. Some of these paths include the following:

- Absorption by plants
- Evaporation from the sun's heating
- Storage in the upper levels of soil
- Storage as groundwater deep in the Earth
- Storage in glaciers and polar regions
- Storage or transport in springs, streams, rivers, and lakes
- Storage in the oceans

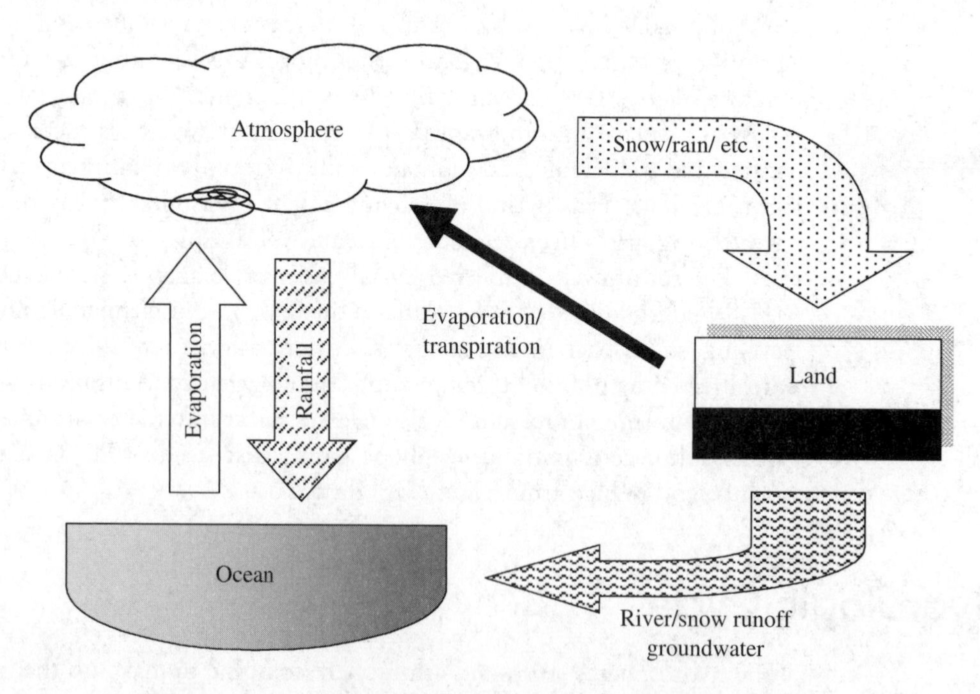

Figure 10.6 The hydrologic cycle is a dynamic system.

When water is stored for any length of time, it resides in a *water reservoir*. A reservoir is a holding area. Nature's reservoirs are oceans, glaciers, polar ice, underground storage (*aquifers*), lakes, rivers, streams, the atmosphere, and the biosphere (within living organisms). Surface water in streams and lakes returns to the atmosphere as a gas through evaporation.

Water held inside plants returns to the atmosphere as a vapor through a biological process called *transpiration*. When plants pull water up through their roots from the soil, use some of the dissolved minerals to grow, and then release the water back through the leaves, the entire cycle is known as *evapotranspiration*. This happens the most during times of high temperatures, wind, dry air, and sunshine. In temperate climates, this is summertime.

When air currents rise into the colder atmospheric layers, water vapor condenses and sticks to tiny particles in the air; this is called *condensation*. When water vapor coats enough particles (dust, pollen, or pollutant), it forms a cloud. When the air is saturated, gravity wins, and water falls as precipitation.

Precipitation can take the form of rain, snow, sleet, or hail depending on the temperature and other atmospheric conditions.

Although the hydrologic cycle balances what goes up with what comes down, in polar regions rain is stored as snow or ice on the ground for several months in winter. In glacial areas, storage extends from years to thousands of years. Then, as temperatures climb in the spring, water is released. When this happens in a short period of time, flooding occurs.

Evaporation

The sun provides energy, which powers evaporation. When water is heated, its molecules get excited and vibrate so much that they break their chemical bonds. Solar energy causes water to evaporate from oceans, lakes, rivers, and streams. Warm air currents scoop water vapor into the atmosphere.

When water changes its form from a liquid to a gas, it is said to *evaporate*.

Because of the huge amount of water in the oceans, it makes sense that roughly 80% of all evaporation comes from the oceans, with 20% coming from inland water and plant transpiration. Wind currents transport water vapor around the world, influencing air moisture worldwide. Hydrologists (scientists who study the Earth's water cycle) estimate that 100 billion gallons of water a year are cycled through this process.

Without the hydrologic cycle, life on Earth would not have developed. Nearly every creation story tells how oceans were formed before continents and their inhabitants. We use water for everything, both internally and externally. Without water, life would not exist on Earth: it is second in importance only to the air we breathe.

Condensation

Water condensation takes place when the air or land temperature changes. Water shifts form when temperatures rise and fall. You see this in the early morning when dew forms on plants.

As water vapor rises, it gets cooler and eventually condenses, sticking to minute particles of dust in the air. Condensation describes water's change from its gaseous form (vapor) into liquid water. Condensation generally takes place in the atmosphere when warm air rises and then cools and loses its ability to cling to water vapor. As a result, extra water vapor condenses to form cloud droplets.

Rainfall differences are affected by a land's topography (shape). For example, mountain topography changes wind patterns, which change precipitation patterns. A rain shadow occurs when warm, moist air is forced to rise over high mountain passages, where it cools and condenses into rainfall. Dry air continues on over the mountains. Depending on climatic conditions, clouds form, winds blow them around the globe, and water vapor is distributed. When clouds can't hold any more moisture, they dump it as snow, rain, or other form of precipitation.

Transport

Next in the hydrologic cycle, transport describes the movement of atmospheric water. Commonly, this water moves from the oceans to the continents. Some of the Earth's moisture transport is visible as clouds, which consist of ice crystals and/or tiny water droplets. Clouds are propelled from one place to another by the jet stream, surface circulations (e.g., land and sea breezes), or other mechanisms. However, a typical 1-kilometer-thick cloud contains only enough water for roughly 1 millimeter of rainfall, whereas the amount of moisture in the atmosphere is usually 10 to 50 times greater.

Transpiration

Another type of evaporation adding to the hydrologic cycle is transpiration. This is a little more complicated. During transpiration (or evapotranspiration), plants and animals release moisture through their pores. This water rises into the atmosphere as vapor.

Transpiration is most easily seen in the winter when you see your breath. When exhaling carbon dioxide and used air, you also release water vapor and heat. Your warm, moist exhalation on a frosty winter morning becomes a small cloud of water vapor.

Transpiration from the leaves and stems of plants is also crucial to the air-scrubbing capability of the hydrologic cycle. Plants absorb groundwater through their roots deep in the soil. Some plants, like corn, have roots a couple of meters in length, while some desert plants have to stretch roots over 20 meters down into the soil. Plants pull water and nutrients up from the soil into their leaves. It is estimated that a healthy, growing plant transpires 5 to 10 times as much water volume as it can hold at one time. This pulling action is driven by water evaporation through small pores in a leaf. Transpiration adds approximately 10% of all evaporating water to the hydrologic cycle.

› Review Questions

Multiple-Choice Questions

1. When calcium carbonate is used to build the shells of sea creatures, it is called

(A) lime
(B) condensation
(C) sublimation
(D) biomineralization
(E) transfiguration

2. By dry weight, approximately what percent of carbon are humans composed of?

(A) 20%
(B) 50%
(C) 65%
(D) 70%
(E) 90%

3. Scientists have studied the carbon cycle in all the following geochemical reservoirs except

(A) the oceans
(B) soil
(C) fossil fuels
(D) the Earth's core
(E) plant life

4. Carbon-containing material from living or non-living sources is called

(A) pyroclastic material
(B) inorganic material
(C) organic material
(D) sublimation
(E) biomineralization

5. When freshwater has high concentrations of calcium and/or magnesium, it is commonly called

(A) calciferous
(B) hard water
(C) soft water
(D) mineralization
(E) lime

6. The residence time of oxygen in the atmosphere is

(A) 12 weeks
(B) 300 years
(C) 1,000 years
(D) 6,000 years
(E) 1 million years

7. When plants pull water from the soil, use the dissolved minerals to grow, and release the water back through the leaves, it is known as

(A) evapotranspiration
(B) respiration
(C) condensation
(D) transport
(E) condensation

8. Plants are known as primary producers of

(A) methane
(B) biomineralization
(C) biomass
(D) water pollution
(E) limestone

9. The idea that matter is neither created, nor destroyed, but recycled through natural cycles is known as

(A) conservation of matter
(B) origin of species
(C) residence time
(D) transconfiguration
(E) respiration

10. All the following are natural water reservoirs except

(A) aquifers
(B) living organisms
(C) streams
(D) limestone
(E) the atmosphere

11. A plant's reaction that captures visible light wavelengths (0.4 to 0.7 μm) and transforms them into chemical energy is known as

(A) sublimation
(B) biomineralization
(C) biomass
(D) evapotranspiration
(E) photosynthesis

12. Organisms that consume litter, debris, and dung are called

(A) carnivores
(B) herbivores
(C) detritivores
(D) parasites
(E) omnivores

13. When ammonia is taken up by plants, dissolved by water, or remains in the soil to be converted to nitrates, it is known as

(A) calcification
(B) residence time
(C) photosynthesis
(D) nitrification
(E) neutralization

14. The location where planetary water is stored for a length of time is called

(A) sink hole
(B) geological cycle
(C) lake
(D) karst
(E) water reservoir

15. Inorganic sulfur is found primarily in rock and minerals as iron pyrite and

(A) gneiss
(B) iron oxide
(C) gypsum
(D) limestone
(E) granite

16. All things that take up space and have mass are known as

(A) sedimentary rock
(B) heavy metals
(C) matter
(D) carnivores
(E) core materials

17. A growing plant transpires up to ___ times as much water volume as it holds at one time.

(A) 2
(B) 10
(C) 18
(D) 22
(E) 25

18. When the temperature of the air or land changes,

(A) condensation of water vapor occurs
(B) populations migrate
(C) leaves change color
(D) snow falls
(E) transpiration is increased

19. When plants absorb carbon dioxide and sunlight to make glucose and build cellular structures, it is known as the

(A) calcium cycle
(B) sulfur cycle
(C) hydrogen cycle
(D) phosphorus cycle
(E) biological carbon cycle

20. Matter exists in three forms, solid, liquid, and

(A) metallic
(B) ice
(C) nuclear
(D) gas
(E) pyrotechnic

› Answers and Explanations

1. D—As these organisms die, their shells fall to the ocean floor and gather or dissolve.

2. B

3. D—Direct sampling is impossible, but computer modeling and analysis have been done.

4. C

5. B—When this water evaporates, a lot of minerals are left behind.

6. D—Although it is a long time, it is much less than nitrogen.

7. A—Water is released to the atmosphere as a vapor from the leaves.

8. C—*Biomass* is a shortened way to describe biological material.

9. A—This principle is also called conservation of mass.

10. D

11. E—Photosynthesis is the main way plants get energy from the sun.

12. C—These organisms consume accumulations of disintegrated material.

13. D

14. E

15. C—A hydrated sulfate of calcium and used for making plaster of Paris.

16. C

17. B

18. A—Water shifts form (e.g., ice, water, vapor) when temperatures rise and fall.

19. E—Carbon plays important structural and processing functions in most life forms.

20. D

Free-Response Questions

1. The geological carbon cycle is intricate with several players. Nature is all about balance. This might be the case with increases in atmospheric carbon dioxide. It may take a long time for the oceans to increase their uptake of carbon dioxide. When the land, sea, and air are overloaded, nature has a tough time keeping up. How does deforestation create imbalances in carbon dioxide levels?

2. The annual rainfall of the state of Washington is more that 450 cm/year, while other areas in the United States get only 20 cm/year or less. High rainfall amounts are found mainly on the western side of the Cascade Mountains, while light rainfall is found on the eastern, or rain shadow, side of the mountain. How do mountains affect rainfall patterns?

3. Limestone is a bedded sedimentary rock made up mostly of calcium carbonate. It's the most important of the carbonate rocks, consisting of sedimentary carbonate mud and calcium-based sand and shells. Limestone is fairly insoluble. However, when plant roots and soil organisms of all sizes give off carbon dioxide, which in turn combines with groundwater, the result is carbonic acid. Carbonic acid dissolves limestone, releasing calcium. How does limestone help proliferate the geological carbon cycle?

Free-Response Answers and Explanations

1. Because the carbon cycle is a closed system, extreme changes in carbon dioxide levels can directly affect weather patterns and adversely affect our environment. Because plant life pulls carbon dioxide from the atmosphere to create glucose and other carbohydrates, helping to keep CO_2 levels in balance, deforestation hinders this invaluable process.
2. Rainfall differences are affected by a land's topography (shape). Like the Himalayas, the Cascade Mountain Range alters precipitation by creating a rain shadow. When warm, moist air is forced to rise due to mountain passages, it cools and condenses creating rainfall, while the dry air continues on over the mountains creating an arid climate, or rain shadow, on the other side. Mountain topography changes wind patterns, which change precipitation patterns.
3. Limestone, which erodes through weathering, falls into oceans creating sedimentary rock that over time is subducted deep into the Earth's crust. There, extreme temperatures and pressures cause the limestone to react with other chemicals through metamorphism, thus freeing carbon dioxide. The carbon dioxide is later released back into the atmosphere through volcanism.

› Rapid Review

- The change between total carbon released to the atmosphere and the total pulled back down governs whether the land is a supplier or reservoir of atmospheric carbon.
- Known as the building block of life, carbon is the foundational element of all organic substances.
- The carbon cycle involves the Earth's atmosphere, fossil fuels, oceans, soil, and plant life of terrestrial ecosystems.
- Precipitation can be rain, snow, sleet, or hail depending on temperature and other atmospheric conditions.
- Transpiration is another type of evaporation in the hydrologic cycle.
- When elements move from one Earth storage form to another, it is known as a geochemical cycle.
- All things that take up space and have mass are known as matter.
- Conservation of matter states that matter is neither created, nor destroyed, but recycled through natural cycles.
- Residence time equals the average amount of time a chemical element (e.g., carbon, calcium, or phosphorus) spends in a geological reservoir or cycle.
- Photosynthesis defines a series of reactions in plants, bacteria, and algae that capture visible light wavelengths (0.4 to 0.7 μm) and transform light energy into chemical energy needed for organic molecules to bond.
- The biological carbon cycle occurs when plants absorb carbon dioxide and sunlight to make glucose and other sugars (carbohydrates) to build cellular structures.
- High concentrations of dissolved calcium and/or magnesium in freshwater cause hard water.
- Water takes a variety of paths and time periods to get back into the atmosphere including
  - Absorption by plants
  - Evaporation from the sun's heating
  - Storage in the upper levels of soil
  - Storage as groundwater deep in the Earth

  - Storage in glaciers and polar regions
  - Storage or transport in springs, streams, rivers, and lakes
  - Storage in the oceans
- Besides geological cycling, sulfur cycles through organisms such as sulfur bacteria that anchor sulfur or release it into the environment.
- Nitrogen (N_2) makes up 79% of the atmosphere, and all life on Earth requires nitrogen-containing compounds (e.g., proteins) to survive.
- At the cellular level, phosphorus is important in energy-transfer reactions.
- The dividing line that separates an area where calcium carbonate dissolves and accumulates is called the lysocline.
- When calcium carbonate is used by marine inhabitants to build shells, it is called biomineralization.

Population Biology and Dynamics

IN THIS CHAPTER

Summary: By understanding the way populations are defined, how they grow, and what their needs are, scientists and policy makers can better plan for the sustainability of resources needed by a population and its ecosystem.

Keywords

✪ Population, ecosystem, exponential growth, carrying capacity, environmental resistance, *r*- and *K*-adapted species, immigration, emigration, mortality, life span, genetic drift

---

Populations

In general, most people think of human populations when they hear the word population. However, to understand a population, you have to think in terms of groupings. A grouping of individuals of the same species located in the same geographical area is known as a *population.* This concept is applied to humans, prairie dogs, or sunflowers. Several populations of the same species in the same area are known as a *community.* When the environment is added into the overall equation, various populations and communities are considered an *ecosystem.* However, plant and animal species are constantly changing, moving, and dying off. These changes are part of the ecological succession described in Chapter 9.

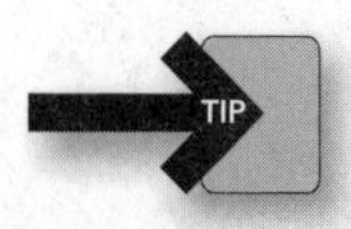

Besides increasing birth rates or germination, populations also grow when organisms enter or *immigrate* into an ecosystem. In the case of seeds, they may be transported by birds

or on the wind. Some animals may fly or swim into a new area, while others are hitchhikers on the fur or feathers of another species or carried in a ship's bilge water.

Internal Population Factors and Life Span

Lots of external factors affect populations. In fact, internal population factors are often controlled by changes in external factors. *Natality,* the germination, cloning, birth, or hatching of new individuals in a population, is affected by external factors (e.g., climate, temperature, moisture, and soil), which determine whether a population will grow or shrink.

Fecundity is the actual capability to reproduce, while *fertility* is a measure of the number of offspring produced. In animal populations, the fecundity of individuals determines a populations' fertility. However, in humans, if two people are physically capable (fecund), it is a matter of choice (with all the birth control options available) whether or not to reproduce.

At the other end of the reproductive spectrum, *mortality* or *death rate* is an internal factor increased or decreased by outside conditions such as extreme heat or cold. Mortality is calculated by dividing the number of individuals that die during a specific time period by the number alive at the start of the time period. Another way of looking at this is in terms of *survivorship,* which describes the number of individuals born at or near the same time, surviving to a specific age.

Life expectancy is the number of years a person is expected to live based on statistical probability. However, life expectancy for a specific person is not set. At birth, a person might be expected, statistically, to live to be 75 years old. However, since many people die at infancy and during childhood, a person who lives to be 74, and is still healthy, will most likely live 10 more years. Except for accidents, he or she will most likely exceed expectancy, rather than die the next year after beating the odds of early death.

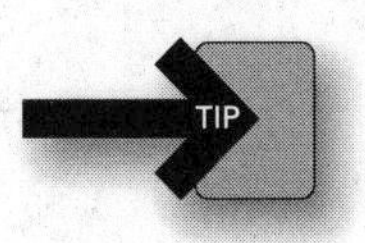

Life span describes the longest interval of time a certain species is estimated to live. Life spans range from a matter of minutes to thousands of years. The maximum human life span is around 120 years. Figure 11.1 shows the four main life span types.

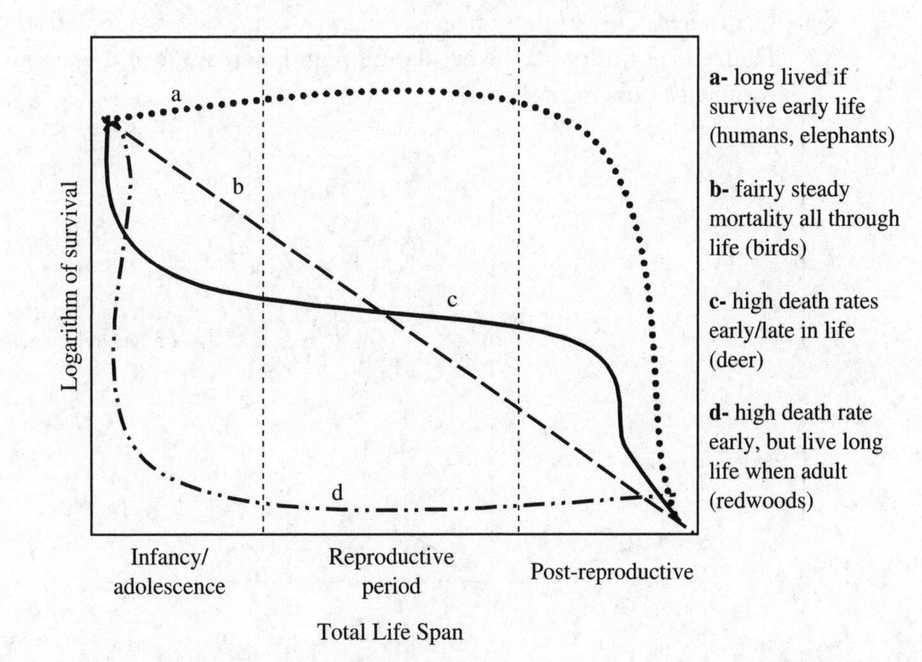

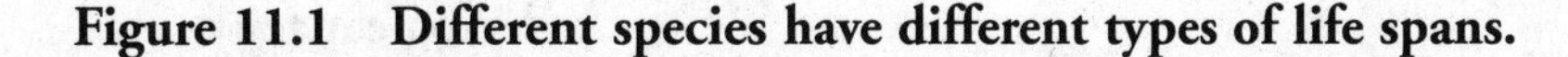

Figure 11.1 Different species have different types of life spans.

Exponential Growth

Sometimes populations experience a growth spurt, which is unrestricted for a time before lack of resources, space, or disease slows or stops it. This is known as *exponential growth.* The mathematical formula is written as the rate of growth (r) times the number of individuals (N), which is equal to the change in the number of individuals (ΔN) in a population over a change in time (Δt).

$$rN = \frac{\Delta N}{\Delta t}$$

The r component describes the average contribution of an individual to population growth. This equation finds a species' *biotic potential.*

Researchers use a common math trick, called the rule of 70, to estimate population doubling. By dividing 70 by the annual percentage growth rate, you get a population's approximate *doubling time* in years. For example, if a population is growing at 20% annually, the population will double in 3.5 years.

Carrying Capacity

Every population experiences rising and falling growth due to environmental factors, social changes, food availability, and disease.

The peak number of individuals of a species supported in a sustainable manner by an ecosystem is called its *carrying capacity.*

At times a population overshoots available resources and death rates are higher than birth rates. This causes a negative growth curve and a population crash generally follows. Figure 11.2 illustrates the oscillating population surge and crash to either side of an environment's carrying capacity.

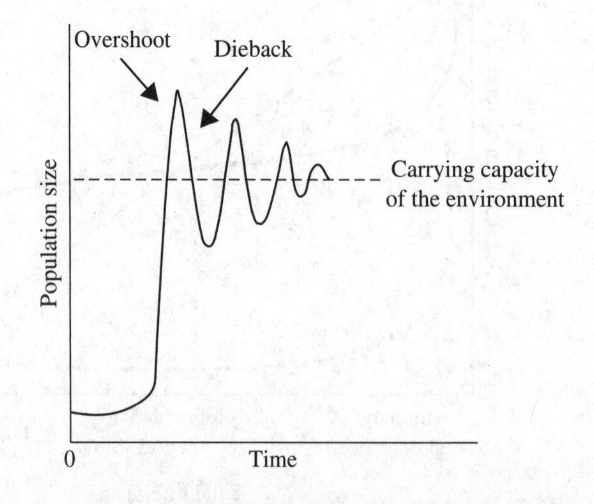

Figure 11.2 Populations increase/decrease with respect to their environment's carrying capacity.

Population Growth

Some populations grow more slowly for internal and external reasons. Internal factors include hormone regulation, maturity, and body size. External growth limitations in the environment include food and habitat accessibility, as well as predator populations. Dense populations increase pressure on resources (e.g., food and water) and overcrowding, which causes increased stress and disease. These population factors are called *density-dependent limitations.* Other population limiters are *density-independent,* such as drought, early frost, fires, hurricanes, floods, earthquakes, and other environmental happenings. Things that lower population density and growth rates are known as *environmental resistance* factors.

A population may grow quickly for a time but then reach its carrying capacity and maintain a stable population size. When the growth rate changes to match local conditions, it is known as *logistic growth.* Logistic growth is described mathematically by the following equation, where K is the environment's carrying capacity.

$$rN\left(1 - \frac{N}{K}\right) = \frac{\Delta N}{\Delta t}$$

The change in population numbers over time ($\Delta N/\Delta t$) is equal to exponential growth over time (r times N) times the carrying capacity (K) of the population size (N). The term $(1 - N/K)$ stands for the relationship between N, a point in time, and K, the number of individuals the environment is able to support. When N is less than K (e.g., 50 compared with 75), then $(1 - N/K)$ is a positive number $(1 - 50/75 = 0.33)$ and slow steady population growth results. However, if N is greater than K, 75 compared with 50, then $(1 - N/K)$ is a negative number $(1 - 75/50 = -0.05)$ and a population's growth exceeds its environmental carrying capacity. Logistic growth describes population drop when it overreaches carrying capacity.

Human Population Change

Humans travel to new locations for family, food, jobs, land, religious freedom, and to avoid war. Population movement from a place is known as *emigration.* When people join a population, it's called *immigration.* Population movement helps researchers predict overall resource requirements, diseases, and other factors important to policy planning. In fact, the U.S. Census Bureau reports an overall drop in the human world population growth rate since 1950 from a high of 2.2% to around 1% today. The projection predicts further decline, with a drop to about 0.4% by 2050.

Human population changes are also related to *replacement birth rate,* or the number of children a couple has to replace them in a population. Two seems the obvious answer, but children die in childhood and some couples don't have children, so statistically the human replacement birth rate can be as high as 2.5.

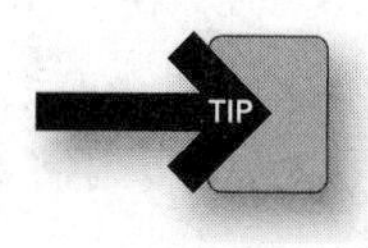

Culture, birth control availability, women's education, public and private retirement programs, child labor needs (e.g., farming), and religious beliefs all impact human fertility rates. Human population growth is also affected by death rates related to (1) mothers dying in childbirth and infant mortality, (2) infectious disease, (3) hazardous work conditions, (4) poor sanitation, (5) clean water supplies, (6) better health care, and (7) better and more available food.

Age–Structure Diagrams

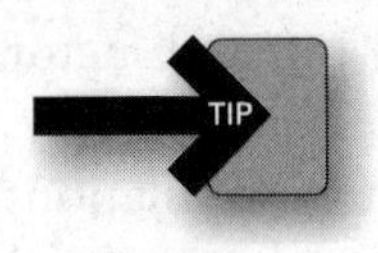

Scientists construct *age–structure diagrams* to study population growth. These diagrams offer a quick way to see what's happening in a specific population and to predict trends. They are often plotted by age (e.g., pre-reproductive 0 to 14 years, reproductive 15 to 45 years, and post-reproductive 46 years and older). Figure 11.3 shows two human populations with different age–structure diagrams and population distributions. Population A has a fairly equal distribution, while population B has many more individuals in the pre-reproductive and reproductive age groups. Population B will have a steep future growth rate, with population A growing more slowly.

Policy makers and researchers also predict population trends based on birth and death rates. They use a *demographic transition model* where zero population growth results from high (↑) birth and death rates or low (↓) birth and death rates. When populations shift from one growth type to the other, it is known as demographic transition. There are generally four demographic types. These include

- *Preindustrial.* Population has a slow growth rate and ↑ birth and ↑ death rates due to difficult living conditions.
- *Transitional.* ↑ birth rates, but better living conditions, allow ↓ death rates and create fast population growth.
- *Industrial.* Population growth is slow with ↓ birth rates and ↓ death rates as seen in developing countries.
- *Postindustrial.* Population nears and reaches zero population growth. Some populations may even dip below the zero growth rate.

Figure 11.4 illustrates a demographic transition model.

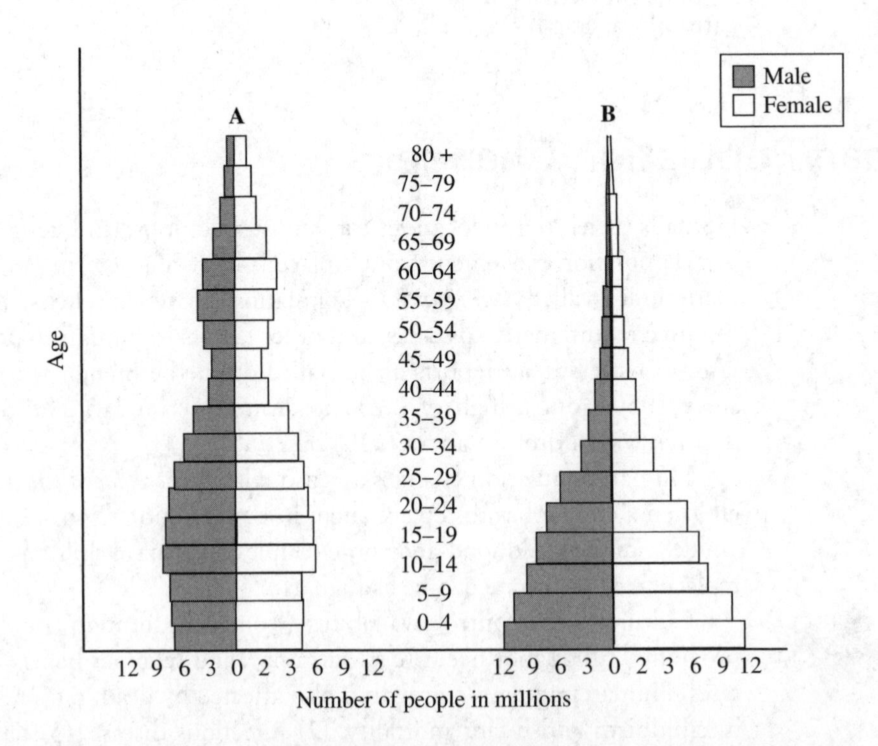

Figure 11.3 Age–structure diagrams allow scientists to study populations and predict trends.

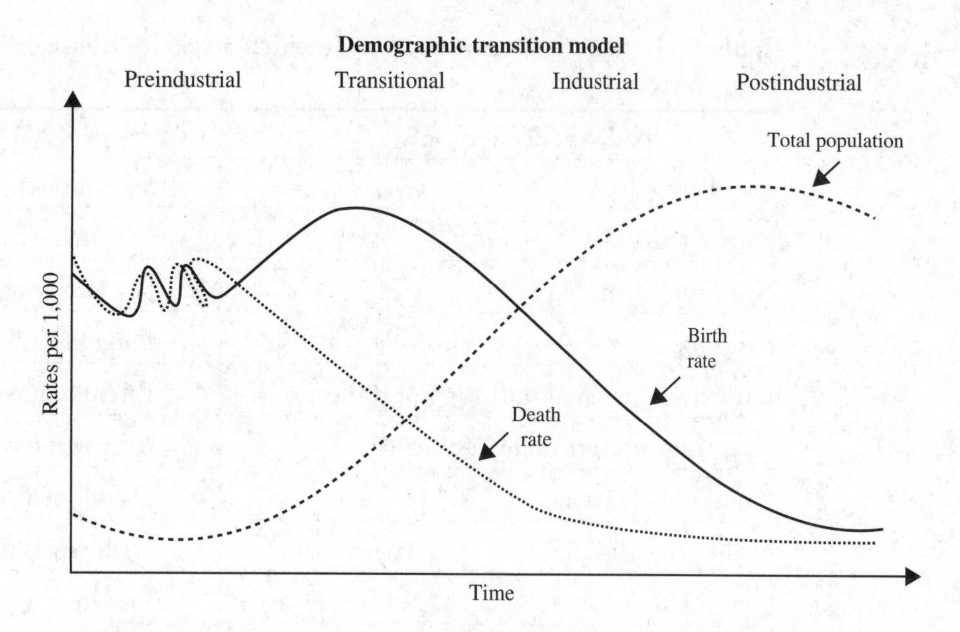

Figure 11.4 Population growth shifts follow four basic demographic transition types.

Population Density

Population density is affected by internal (within the population, like survivorship) factors, as well as external (e.g., environmental) factors. These can be further broken down into *biotic* (e.g., living organisms like the plague) or *abiotic* factors (related to nonliving elements like drought). How these various influences affect a population is also *density dependent* or *density independent.* For example, the plague has a higher mortality rate and spreads faster in a dense population than in a population where individuals are farther apart. Drought and volcanic eruptions are density independent, affecting populations in the same area in the same way whether they are densely packed or not.

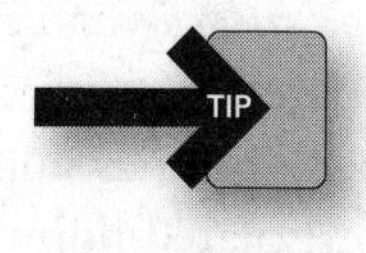

Density-dependent factors are less random and affect populations more as density increases. For example, if habitat is lost or food eaten, competition for resources grows. Fighting between and within species, lack of females, stress, low birth rates, and disease all increase. Overcrowding forces predators closer to their prey. When this happens, old and sick individuals are killed by predators, but the herd becomes stronger.

Species Adaptation

Species can also be categorized in the ways in which they adapt to their environment. The first, known as *r-adapted species,* produce lots of offspring and have a high rate of growth (rN). These opportunistic species are generally lower on the food web and use great numbers of offspring to ensure species survival. Weeds, such as dandelions, which spread almost overnight in a lawn, fit into this category.

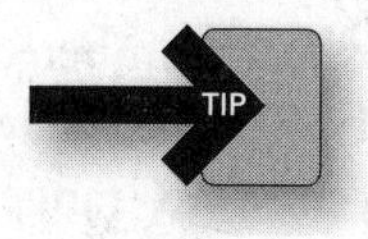

Long-lived species, which produce slower maturing and fewer offspring, like elephants (e.g., reaching reproductive maturity around age 20 and giving birth only every 4 to 5 years), maintain a generally stable population with the carrying capacity (K) of their environment. Elephants, then, are called *K-adapted species,* and tend to live 60 to 70 years if not heavily impacted by density-independent factors such as habitat loss or heavy poaching. Adaptation strategies are detailed in Table 11.1.

Table 11.1 The way and speed with which a species adapts is dependent on many factors.

r-ADAPTED SPECIES	K-ADAPTED SPECIES
Rapid growth	Slow growth
Early maturity	Late maturity
Numerous offspring	Few offspring
Short life	Long life
Little to no parental nurturing or protection	Parental nurturing and protection
Adaptation to varied environment	Adaptation to stable environment
Pioneers, colonizers	Established conditions
Niche generalists	Niche specialists
Prey	Predators
Affected by internal factors	Affected by external factors
Low on food web	High on food web

Conservation Biology

The old saying "You can't fool Mother Nature" seems to be true when predicting, protecting, and assessing species populations. Conservation biologists are only beginning to understand all the physiological and environmental factors that regulate populations. Add in human pressures from overfishing, overhunting, and habitat destruction, and you begin to see the complexity of the problem.

However, some geographically isolated populations are classified as *island biogeographic.* Geographic isolation has pros and cons for a population. Things that barely disturb a diverse, unrestrained population (e.g., disproportion between the number of males and females) can annihilate an isolated population. This is particularly true for isolated populations when genetic diversity narrows due to few or no new individuals coming into the population. For recessive traits to be balanced out, a certain number of individuals with dominant and healthy traits need to reproduce. In a big population, new genes are always circulating among mating individuals and the population remains healthy.

However, with limited numbers of breeding pairs on an island, a mutation would be passed on through subsequent generations unequally. When a species is isolated long enough, loss of genetic diversity may affect reproduction, adaptability, and species survival. This founder effect, mentioned in Chapter 9, plays a part in genetic drift.

Conservation biologists study isolation, genetic drift, and founder effect to figure out the minimum viable population size, which is the lowest number of individuals needed for a species' continued survival. For example, in North America, the grizzly bear numbered around 100,000 in 1800 and occupied an area from the Midwest to the Pacific Ocean. Today, fewer than 1,000 grizzlies occupy six areas on less than 1% of their previous range. Yellowstone National Park in Wyoming has around 200 grizzly bears. Conservationists aren't sure if this low number can sustain genetic viability and avoid problems if the

Yellowstone population becomes even more isolated through resource and habitat loss. Even introducing a few genetically diverse bears every few generations helps increase population viability.

Impacts of Population Growth

Because of population growth, humans have impacted the Earth's resources more than any other species. The availability of clean water, sanitation, improved medical treatments, and advanced food production methods have all lengthened the average human life span and added to global population growth.

Historically, many countries have been able to feed their populations with local resources. However, population growth increases food demand and the demands on these resources. Some regions (e.g., Latin America), previously self sufficient, must now import grain and other products to feed their growing populations.

Land Use

Land overuse results from economic circumstances, poor land laws, and cultural customs. Some people exploit land resources for their own gain with little thought for the land or neighboring areas. Some people in poverty have little choice but to overuse their meager resources, even to the extent of wearing out the land. Trade and exploitation of a country's natural resources often leaves land restoration in the hands of local people without any funding.

Wars and national emergencies also destroy rich land by overburdening it with refugees and other displaced people. Natural disasters like floods and droughts can do the same thing. All this limits environmental resources needed for regional populations to maintain population growth and health.

Resource Distribution

Some charities (e.g., local food banks and national agencies) exist that redistribute food. They get food from growers, food processors, and distributors, and then make the food available at low prices to needy people.

Internationally, groups like the World Trade Organization have policies controlling international trade. However, developing countries are often at a disadvantage since they have used many of their resources and have nothing to export except labor. In order to secure work for their population, these companies often bid for international labor contracts at very low prices. This does little to pull them out of a continuing cycle of poverty.

Population Control

Efforts to control population include birth control and education (especially of women). Some countries, such as China with extreme overpopulation over 1.4 billion people, limit population growth by enacting laws (e.g., one child only per couple) to raise everyone's standard of living and prevent hunger and poverty resulting from unrestrained population growth. China's One Child Policy, though controversial in the West due to female gender discrimination, has resulted in a decrease in the population growth of that country by approximately 250 million.

› Review Questions

Multiple-Choice Questions

1. Change in gene frequency is known as
   (A) mitosis
   (B) genetic drift
   (C) hybridization
   (D) genetic optimization
   (E) founder effect

2. Which term is used to describe the longest time interval a certain species is estimated to live?
   (A) Mortality
   (B) Nutrition
   (C) Decades
   (D) Life span
   (E) Fertility

3. When populations experience an unrestricted overshoot before limited resources, space, or disease cause a dieback, it is known as
   (A) logarithmic growth
   (B) a *p* curve
   (C) exponential growth
   (D) proven reserves
   (E) biotic exponential

4. Fecundity is the actual capability to reproduce, while number of offspring produced is a measure of
   (A) ecological succession
   (B) natality
   (C) survivorship
   (D) fertility
   (E) niche development

5. The rule of 70 is used to estimate
   (A) entrepreneurial species
   (B) population doubling
   (C) density-dependent species
   (D) infant development
   (E) potential of a new Starbucks in a neighborhood

6. Movement out of a population is known as
   (A) delinquency
   (B) mortality
   (C) emigration
   (D) suburban sprawl
   (E) immigration

7. The number of years an individual is statistically likely to live is called his or her
   (A) species duration
   (B) natality
   (C) life expectancy
   (D) mortality
   (E) genetic drift

8. Isolation, genetic drift, and founder effect help biologists figure out
   (A) species adaptability
   (B) succession level
   (C) mutualism
   (D) maximum viable population size
   (E) minimum viable population size

9. Biotic potential is defined by which of the following equations?
   (A) $rN = \Delta N/\Delta t$
   (B) $rN = \frac{1}{2}\Delta N - t$
   (C) $rN = \Delta t/\Delta N$
   (D) $rN = \Delta t/N/K$
   (E) $rN(1 - N/K) = \Delta N/\Delta t$

10. When a species produces lots of offspring and has a high rate of growth (rN), it is known as a
    (A) pioneer species
    (B) *K*-adapted species
    (C) long-lived species
    (D) *r*-adapted species
    (E) foundational species

11. Preindustrial populations usually have slow growth rates and
    (A) low economic potential
    (B) ↑ birth and ↑ death rates
    (C) low geographic resources
    (D) ↓ birth and ↑ death rates
    (E) ↓ birth and ↓ death rates

12. Biotic and abiotic factors greatly affect
    (A) population density
    (B) niche development
    (C) fecundity
    (D) education levels
    (E) genetic drift

13. Human population control

(A) has never worked
(B) has reaped substantial benefits in China
(C) is impossible in developed nations
(D) is not needed with a slowing global population
(E) is not possible with today's technology

14. A group of individuals of the same species in the same geographical area is known as a

(A) herd
(B) niche
(C) population
(D) system
(E) flock

15. Exponential human population growth is affected by all the following factors except

(A) infectious disease
(B) clean water supplies
(C) hazardous work conditions
(D) increased television time
(E) better and more available food

16. The germination, cloning, birth, or hatching of individuals in a population is known as

(A) life span
(B) niche
(C) mortality
(D) genetic drift
(E) natality

17. The number of individuals born at or near the same time and living to a specific age is called

(A) survivorship
(B) natality
(C) mortality
(D) fecundity
(E) immigration

18. When a species is isolated long enough, loss of genetic diversity may affect reproduction, adaptability, and species survival. This is known as

(A) *r*-adapted species
(B) *K*-adapted species
(C) founder effect
(D) logistic growth
(E) demographic transition

19. When a population's growth rate changes to match local conditions, it is known as

(A) exponential growth
(B) density-independent growth
(C) natality
(D) logistic growth
(E) pioneer growth

› Answers and Explanations

1. **B**—New genes changing and circulating in a population keep the species healthy.

2. **D**—The longest recorded human life span is 122 years.

3. **C**—The unrestricted growth rate is expressed as a fraction or exponent by which the starting population is multiplied.

4. **D**—Rabbits are more fertile (many babies often) than elephants (one baby infrequently).

5. **B**—By dividing 70 by the annual percent rate of growth, you get a population's estimated doubling time.

6. **C**—People coming into an established population are called immigrants.

7. **C**—The life expectancy for a specific person is not set but affected by many factors.

8. **E**—Introducing genetically diverse individuals increases population viability.

9. **A**

10. **D**—These species are lower on the food web and produce great numbers of offspring.

11. **B**—This is due to hard living conditions and primitive medical care.

12. **A**—Fighting between species, lack of females, stress, low birth rates, and disease, along with tornadoes, flash floods, and drought, affect population.

13. **B**—By slowing population growth, China can better meet its food needs.

14. **C**

15. **D**

16. **E**—It is affected by external factors such as climate and temperature.

17. **A**—Survivorship is an internal population density factor.

18. **C**—The isolated population often has a tough time adapting to new conditions.

19. **D**—This is known as logistic growth and is equal to a population's carrying capacity.

Free-Response Questions

1. Most people blame desertification on overpopulation. However, it's possible for large populations to practice good conservation and avoid desertification. It is said the United States could feed most of the world with all the grain produced in the Midwest. However, the Ogallala Aquifer, which provides most of the water for irrigation there, has been overmined for years. Crop irrigation there might not be possible in another 25 to 40 years.

Desertification is a complex problem since a decrease in population can also cause desertification because there are less people to take care of the land. In some countries, when young people in villages go into the city to find work, their aging parents can't keep up with the land's needs and cultivation.

(a) What are some conservation and land management methods large populations can use to prevent desertification?
(b) Describe three impacts (other than desertification) human population growth has had on the environment.
(c) How could overmining of the Ogallala Aquifer change the carrying capacity of the "grain belt"?

Free-Response Answers and Explanations

1.

a. Rotational grazing can be used to help increase the health of grasslands by reducing the stress cattle place on the environment through overgrazing, while maximizing the positive effects (e.g., fertilization) cattle have on grasslands. Proper forestry and irrigation techniques can also help, as well as cultural and lifestyle changes to reduce dependence on traditional land use.

b. Human population growth has affected the world in ways far greater than that of any other species. Redistribution (or lack of) of resources (e.g., land, water, oil, and other fossil fuels) according to population needs has reshaped the environment in countless ways, from the building of roadways to war and famine. Our agricultural practices have led to increased life spans, which in turn continuously limit the amount of available resources. Many countries and regions, once self-sufficient (e.g., Latin America) must now import resources to sustain their growing populations.

c. The huge loss of water for irrigation due to overmining the Ogallala Aquifer would drastically reduce the region's carrying capacity. By "living beyond their means," local populations and those benefiting from grain sales have created a possible future where grain shortages could lead to massive socioeconomic and environmental changes.

❯ Rapid Review

- Exponential growth happens when populations experience an unrestricted growth overshoot before limited resources and space or disease cause slowing.
- Logistics growth describes population drop when it overreaches carrying capacity.
- A group of individuals of the same species located in the same geographic area is known as a population.
- When a species is isolated long enough, loss of genetic diversity may affect reproduction, adaptability, and species survival. This is known as the founder effect.
- Fertility is only half the answer to exponential human population growth. The other half comes from the drop in death rates related to (1) mothers dying in childbirth and infant mortality, (2) infectious disease, (3) hazardous work conditions, (4) poor sanitation, (5) clean water supplies, (6) better health care, and (7) better and more available food.
- Natality (i.e., germination, cloning, birth, or hatching of new individuals in a population) is affected by external factors (e.g., climate, temperature, moisture, and soil), which determine whether a population will grow or shrink.
- Fecundity is the actual capability to reproduce, while fertility is a measure of the number of offspring produced.
- Life span describes the longest interval of time that a certain species is estimated to live.
- Life expectancy is the likely number of years an individual is expected to live based on statistical probability.
- Mortality is calculated by dividing the number of individuals that die during a specific time period by the number alive at the start of that same time period.
- Survivorship describes the number of individuals born at or near the same time and living to a specific age.

- The first species to inhabit an area is called a pioneer species.
- External growth limitations in the environment include food and habitat accessibility, as well as predator numbers.
- Mortality or death rate is an internal factor that can be increased or decreased by external conditions such as extreme heat or cold.
- Movement out of or away from a population is known as emigration.
- When individuals join a population, it is called immigration.
- Population limiters, such as drought, early frost, fires, hurricanes, floods, earthquakes, and other environmental happenings, are density independent.
- Things that lower population density and growth rates are called environmental resistance factors.
- The r component in $rN = \Delta N/\Delta t$ describes the average contribution of an individual to population growth. This equation finds a species' *biotic potential.*
- The rule of 70 is used to estimate population doubling. By dividing 70 by the annual percent rate of growth, you get a population's approximate doubling time in years.
- Logistic growth is described mathematically by $rN(1 - N/K) = \Delta N/\Delta t$, where K is an environment's carrying capacity.
- Conservation biologists study isolation, genetic drift, and the founder effect to find the minimum viable population size or lowest number of individuals for a species' survival.

Agriculture and Aquaculture

IN THIS CHAPTER

Summary: Cultivatable land is shrinking as more and more is used for housing, businesses, and other societal needs. Developed and developing countries must feed growing populations in sustainable ways by using modern methods to produce high yields of nutrient-rich and high-protein food.

Keywords

✪ Malnourishment, aquaculture, bycatch, target species, monoculture, polyculture, nonnative species, green revolution, genetic engineering, pesticide, biocide

Human Nutrition

Human beings have specific nutritional requirements needed to grow and thrive. These include proteins, vitamins, and minerals. A person can get enough caloric energy from food, but if the right nutrients aren't present in the food, the person becomes malnourished. *Malnourishment* is a nutritional imbalance caused by the lack of important dietary elements or the inability to absorb essential nutrients from food. The United Nations' Food and Agriculture Organization (FAO) estimates nearly three billion people suffer from protein, vitamin, and mineral deficiencies.

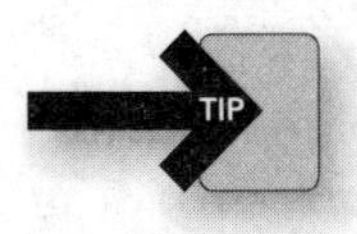

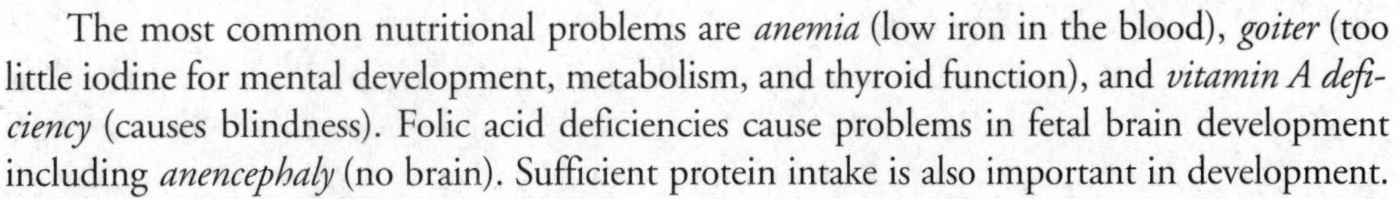

The most common nutritional problems are *anemia* (low iron in the blood), *goiter* (too little iodine for mental development, metabolism, and thyroid function), and *vitamin A deficiency* (causes blindness). Folic acid deficiencies cause problems in fetal brain development including *anencephaly* (no brain). Sufficient protein intake is also important in development.

However, many people in developed countries eat too much protein, salt, sugar, and fat, but not enough fiber. The result is unhealthy weight gain and *obesity,* defined as more than 30 pounds over ideal weight for a person's height and age.

The U.S. Surgeon General reported that in the past decade, 40% of Americans were overweight, compared to 69% today. In fact, over one-third of the total population is obese. The United States is not the only country with weight problems. Unfortunately, obesity is stretching around the world to developed and developing nations alike as more countries mimic Western diets. With this rise, the rates of heart attack, stroke, diabetes, and depression are also increasing.

To combat this unhealthy rise, scientists have reconsidered what makes up a healthy diet. The four basic food groups of meat, dairy, fats, and sweets have been sidelined to a much smaller percentage of the daily diet. Modern food pyramids advise larger portions of nuts, beans, fruit, vegetables, plant oils, and whole grains as foundational components. Processed bread, cereal, rice, and pasta should be eaten very sparingly. This recommendation is combined with a recommendation for regular exercise.

Agriculture

Types of Agriculture

The three main grain crops eaten by humans are wheat, rice, and maize (called corn in the United States). Combined, 1,900 million metric tons are grown annually across the globe. In fact, wheat and rice make up 60% of the calories consumed by humans.

Potatoes, barley, rye, and oats, grown at higher altitudes are important in northern Europe and Asia. Roots and tubers, grown in equatorial climates, feed people in the Amazon, Africa, and the South Pacific.

Fruits and vegetables are important because they contribute high levels of vitamins, fiber, carbohydrates, and minerals needed for development and overall health.

Although thousands of plant species have been grown for food over time, only 16 species are currently cultivated. Which plants are grown in an area is determined by climate, pests, soil, and other factors. These plant species include winged beans, which grow well in hot humid climates, and tricale, a hybrid of wheat and rye that flourishes in sandy, infertile soil.

The huge increases in yield obtained through modern cultivation techniques has become known as the *green revolution.* These increases have allowed food crop production to keep pace with worldwide gains in human population.

Genetic Engineering and Crop Production

A hundred years ago, crops were pollinated by wind and insects. The average wheat yield was around 25 bushels per acre. Today it is a different story. By selectively transferring genetic material to make plants stronger, more drought tolerant, and resistant to insects, modern wheat yields average between 130 and 250 bushels per acre. Plant hybrids such as dwarf high-yielding wheat and rice raise production by 3 to 4 times the normal rate.

Genetic engineering is the bioprocess of taking DNA (blueprint proteins) from one species and splicing it into the genetic material of another.

Genetic cultivation methods greatly increase the quality and amount of food produced on the land. When DNA is combined in ways to augment certain characteristics in a species' natural growth, the result is known as a *genetically modified organism* (GMO).

The good thing about GMOs is that they can be created to thrive in environments with frost, low rainfall, high salt, and low nutrients, and even disease resistant. Some plants are designed to produce their own natural pesticides and eliminate the need for toxic chemicals. Some allergens can even be removed from plant strains.

One criticism about genetic engineering is that new strains of superweeds, toxic plant by-products, or other harmful consequences of human tinkering with nature will upset natural ecosystems. If GMOs cross with wild species or produce unwanted resistance to pesticides, later generations may pay the price. One way to limit this problem is to plant some field sections with non-GMO strains to keep insects from developing resistance to antipest strains.

Currently, about 82% of soybeans, 71% of cotton, and 25% of all corn in the United States are GMOs. In fact, it has been estimated that over 60% of all processed food in America has GMO ingredients.

Irrigation

Another agricultural method that has come under scrutiny is water use and distribution. Many farmers use flood, ditch, or sprinkler irrigations, which allow for massive evaporation from the land and crops. More is lost through runoff.

In countries with few water resources, conservation is crucial. Watering plants directly without water logging the soil is important. One method uses *drip irrigation* to provide water directly to thirsty plant roots.

The additional cost of maintaining pumps, canals, dams, and reservoirs is another reason for water conservation and better irrigation methods. In some regions, farmers have overpumped underground reservoirs and lowered the water table to below accessible levels. Now they want government funds for canals and other infrastructure to divert river water for their crops. Opponents point out the impacts of river diversion on fishing, hunting, and other commercial and recreational water use. Since pumping irrigation water is common, these kind of decisions have broad application to worldwide agriculture.

Pest Control

A *biological pest* lowers the quantity, quality, or value of resources. Most people equate pests with insects, since mosquitoes, cockroaches, gnats, and locusts affect humans and crops negatively. In fact, insects make up roughly 75% of all species on Earth. Most insects, however, are harmless to humans. Out of the roughly 1 million species of insects described so far, no more than 1,000 are considered serious pests.

Types of Pesticides

Historically, humans have controlled pests with smoke, salt, sand, and insect-repelling plants. The ancient Sumerians used sulfur to control insects and mites 5,000 years ago. Greeks and Romans used oils, ash, lime, and other compounds to protect themselves, their crops, and their animals from pests.

Modern pest control is performed by eliminating pests' living conditions, such as swamps. *Pesticides* are chemicals that kill or repel certain pests, while *biocides* kill a broader range of organisms. In addition to their use for pest control, biocides are also found in adhesives, paints, leathers, petroleum products, and plastics, as well as recreational and industrial water treatment in totals of around 300 million pounds per year. This large amount accounts for less than 10% of the total U.S. yearly pesticide use.

Around 80% of all pesticides in the United States are used in agriculture or food storage and shipping. Chemicals, which kill specific targets, are named accordingly. These are *insecticides* (kill insects), *herbicides* (kill plants and weeds), *acaricides* (kill spider mites and ticks), *rodenticides* (kill rats and mice), and *fungicides* (kill molds and fungus). Pesticides are also described by application methods such as fumigation (gassing) to treat wood against insects. Pesticides are categorized by chemical structure. These are inorganic, natural organic,

Table 12.1 Pesticides use many different chemicals to eliminate a variety of pests.

PESTICIDE TYPE	CHEMICALS
Inorganic	Mercury, arsenic
Natural organic	Nicotine, rotenone, turpentine, phenols, aromatic oils
Fumigants	Carbon tetrachloride, carbon disulfide, ethylene dichloride, methylene bromide
Chlorinated hydrocarbons	DDT, chlordane, aldrin, dieldrin, toxaphene, lindane, paradichlorobenzene (mothballs)
Organophosphates	Parathion, malathion, dichlorvos, chlorpyrifos, dimethyldichlorovinylphosphate (DDVP)
Carbamates	Urethane, carbaryl (Sevin), aldicarb (Temik), carbofuran (Baygon), Mirex
Microbial/biological agents	Bacteria (*Bacillus thuringiensis*), ladybugs, lacewings, wasps

fumigant, chlorinated hydrocarbons, organophosphates, carbamates, and biological agents. Table 12.1 shows the makeup of these various pesticides.

Costs and Benefits of Pesticide Use

Early environmentalists warned of random pesticide use as early as the 1950s, but it was not until the 1962 publication of the book, *Silent Spring,* by marine biologist Rachel Carson that environmental contamination came into broader public view. She pointed out the hazards of careless application and spraying of herbicides and pesticides.

One such pesticide hazard became apparent with DDT's (dichlorodiphenyltrichloroethane) impact on predatory nontarget bird species 50 years ago. Known as an excellent broad-spectrum insecticide, DDT built up (*bioaccumulation*) in the tissues of lower organisms like zooplankton (0.04 ppm) until it was stored in toxic levels (25 ppm) in the tissues of predatory birds (falcons and eagles) eating poisoned rodents and other small animals. This buildup in the highest members of a food chain is known as *biomagnification.*

Pesticides often poison nontarget species. Pesticide use has been directly linked to mass killings of migrating birds (robins in 1972), and the decline of migrating Atlantic salmon in Canada due to 4-nonylphenol, a strong hormone disrupter.

Pesticide resistance is also a problem since every target species has genetically diverse individuals resistant to pesticides. Since these resistant members of the targeted population then reproduce resistant generations, the population eventually resurges or rebounds from the pesticide hit. Farmers must use increasingly more pesticide to get the same crop protection. The pesticide use becomes an upwardly spiraling pesticide treadmill with farms hesitant to get off due to income loss from pest damage. At least 1,000 insect species and 550 weed and plant pathogens globally have developed chemical resistance.

Because of their stability, high solubility, and high toxicity, pesticides often have a big environmental impact. They accumulate in water and soil for years, and their effects are long lasting. Pesticides also evaporate from contaminated areas and then rain down on far distant areas.

Pesticides are often stored in the fat of animals. Dangerously high levels become concentrated and can cause birth defects and death in predators and prey alike. Species such as polar bears, porpoises, trout, eagles, whales, and humans at the top of the food chain suffer the most.

Pesticide Management

Since insects are increasingly more resistant to a single strategy, scientists and ecologists have come up with *integrated pest management.* This includes flexible and different application methods to kill insects and weeds. For example, hormones, which stop larvae growth, combined with mechanical vacuuming of adult bugs from plants has been successful. Breeding hybrid plants that are resistant to insects also lowers pesticide use. Careful monitoring of economic thresholds between pest damage and pesticide cost and application time is catching on as well.

Trap crops, small plots planted before the rest of the field, are sprayed heavily with pesticides trapping and killing insects that would harm the main crop. The trap crop is destroyed to protect farm workers and consumers from the concentrated pesticide. The rest of the field then grows relatively pest and pesticide free.

Relevant Laws

In the United States, three agencies are responsible for regulating pesticides used on food crops: the Environmental Protection Agency (EPA), the Food and Drug Administration (FDA), and the Department of Agriculture (USDA). The EPA controls the sale and use of pesticides under the Federal Insecticide, Fungicide, and Rodenticide Act, which calls for the registration (licensing) of all pesticide products. The EPA studies risks to humans and the environment from pesticides. Under the Federal Food, Drug, and Cosmetic Act, the EPA sets acceptable pesticide level limits in foods sold in the United States regardless of where it was grown. The FDA and USDA enforce use and allowable levels regulated by the EPA and have the authority to destroy any shipments found to be in violation of EPA limits.

In 1996, Congress passed the Food Quality Protection Act. This allows the EPA to set limits on combined pesticide exposures. This act determined that by 2006 the EPA had to reassess allowable pesticide levels on food and include a safety factor of 10 where complete information was not available on pesticide levels affecting children.

This Act also led the EPA to examine *inert* pesticide ingredients for the first time. These compounds, usually not listed on a label, are used to dilute or transport active chemicals. Of the 2,500 substances used for this purpose, over 650 have been identified as hazardous by local, state, and federal agencies. Over 50% have been identified as carcinogens, occupational hazards, and air and water pollutants. One example, naphthalene, is listed as a hazardous pollutant under both the Clean Air and Clean Water Acts.

Sustainable Agriculture

No one wants to use chemicals that cause cancer or birth defects, or harm nature. However, countries all over the world must feed their growing populations and produce income-generating crops. The trick is to do it in a sustainable way. When farmers grow nutritious crops while stopping harmful past practices, it is called *sustainable agriculture, regenerative farming,* or *agroecology.* Science advancements as well as past mechanical practices may be the answer.

Soil conservation and care is the best place to start by limiting erosion. In Asia, rice cultivation is carefully nurtured with organic material, keeping it healthy year after year. Topography methods such as contour plowing are replacing traditional planting techniques and substantially reducing water runoff and erosion. Mulch and ground cover species are also beneficial to the bottom line. Organic farms allow animals such as cows, chickens, and pigs to roam and eat native grasses and seeds. This produces healthier animals and a source of odorless organic fertilizer. Veterinarian and feed costs on these farms, along with the reduced need for workers, are down.

Marine Resources

The importance of marine resources to humans is obvious. Fish and shellfish give us a valuable source of food protein and livelihood for many in the seafood industry. Other marine resources meet the needs of the general population and provide jobs. The question is, "To what extent can the oceans continue to meet human needs?"

Overfishing

Fisheries in many countries remain unregulated and inadequately studied. Even where research has been done, political and economic pressures override scientists' recommendations. In fact, most fisheries, close to coasts impacted by human activities, have been overfished. This is a major problem with ocean fisheries worldwide.

Since over 15% of all animal protein for humans comes from seafood, the fishing industry always has a market for its catch. To meet this demand, technology has been used to increase yield. Unfortunately, too many boats using efficient technology have impacted three-quarters of the world's edible ocean fish, crustaceans, and mollusks according to United Nations' estimates. Conservation must be observed by all seafaring countries to have any hope of sustaining declining fish populations.

Bycatch

When nontarget fish species or other marine animals are caught in nets or through other fishing methods, they are called *bycatch.* Accidental catching and killing of nontarget species, like nonmarketable "trash" fish and marine mammals (dolphins, sharks, etc.), have also impacted ocean ecosystems. In 2012, based on a new fishery-by-fishery approach, the Food and Agriculture Organization (FAO) of the United Nations estimated that for the years 1992–2001, an average of 7.3 million metric tons of fish were discarded as bycatch by commercial fishermen. This amount is roughly equal to one-third of the total yearly global catch. This does not include the unintended killing of several hundred thousand sea turtles and marine mammals each year. Fishing methods such as *bottom trawling, gill netting,* and *longlining* are especially subject to bycatch. Shrimp trawlers catch and throw away an estimated 9 pounds of nontarget species for every 1 pound of shrimp caught.

As the public finds out about these problems, funding becomes available for research and development of new technologies. For example, when it was learned that thousands of dolphins were caught in tuna nets and drowned before they could be released, new escapable nets were designed. Public demand for "safer" tuna grew, and those companies refusing to change saw sinking sales.

Nonnative (Alien) Species

When plants and animals found in one part of the world, with its own natural predators, diseases, and ecosystem controls, are transported to a far distant location, they are known as *nonnative* or *alien* species. In their new location, they increase, die, or something in between.

> Plants and animals transplanted geographically to a formerly unknown area are known as *nonnative (alien) species.*

Many times, a new species can't live in a new environment and fails to prosper. Other times, however, the new species grows, crowds out native species, and destroys the local ecosystem.

In the ocean, alien species are introduced through a ship's ballast water. When ships fill their ballast tanks to adjust their stability in the water, they suck in the local marine inhabitants. Later after emptying the ballast water elsewhere, stowaway passengers are expelled. Scientists think as many as 3,000 alien species are transported around the world in ships daily.

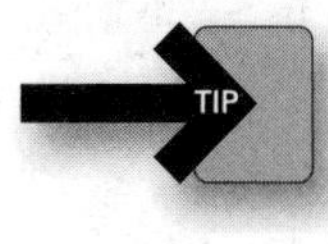

Nonnative species affect coastal waters more than the open ocean. In coastal zones, where transplanted species are often biologically and ecologically distinct, alien species have a greater impact. Species transported from tropical regions of the Atlantic to the Pacific are frequently even more disruptive because there is little natural exchange between these oceans. Transplants between the Indian Ocean and the Atlantic Ocean are not all that different and do not have as serious effects since these oceans are connected.

Because of worldwide shipping, however, the impact of alien species has increased dramatically. Global population growth has forced many nations to bring in food and other products from trade partners. Many countries, formerly self-reliant, now depend on imported goods.

This shipping increase has caused a huge increase in the introduction of marine alien species into new ecosystems incapable of coping with them. For example, in San Francisco Bay, California, the ecosystem has been overwhelmed by many new alien species. Currently there are over 200 different alien species living in San Francisco Bay, severely impacting native species and the overall ecosystem. Although alien species have not completely wiped out native species, the Bay's natural processes have changed a lot.

Besides changing native marine ecosystems, alien species can also affect humans. Zebra mussels, an introduced species in the Great Lakes, have done everything from disturb native species to clogging pipes. However, on the plus side, they increased toxin filtration from the water.

Aquaculture

Growing marine or aquatic species in net pens or tanks is known as *aquaculture.* Today, nearly 50% of the world's seafood comes from fish farms or aquaculture. Fish are raised in ponds that take relatively little land and may not be rich enough to grow other food crops.

However, aquaculture is not without problems. Raising high-value carnivorous species such as salmon and sea bass can reduce the number of wild species, since they are often used to start stock populations. Worldwide, thousands of hectares of mangrove forests and wetlands are destroyed to build fish ponds. These natural environments are essential as nurseries for many different marine species.

Additional problems arise from aquaculture pens anchored near shore where disease, escape of exotic species, and pollution from feces, excess food, and antibiotics endanger native species. This is not a problem with small family operations, but countries with huge aquaculture pens greatly intensify these environmental problems.

Mixed Species

Raising different fish species together (*polyculture*), such as the grass carp (eats plants) and silver carp (filter feeder of small organisms), reduces many aquaculture problems caused by raising only one species (*monoculture*). The fish farm species are healthier and do not deplete the same food supply. Because of these benefits, polyculture farming yields are often 50% higher per hectare than monoculture aquaculture farming.

› Review Questions

Multiple-Choice Questions

1. Zebra mussels are

(A) the largest of the mussels
(B) actually more spotted than striped
(C) an introduced alien species in the Great Lakes
(D) related to zebra clams
(E) a native species

2. Growing marine or aquatic species in net pens or tanks is known as

(A) cross cultivation
(B) a risk-free industry
(C) bioculture
(D) an easy source of iodine
(E) aquaculture

3. Species transported from tropical regions of the Atlantic to the Pacific

(A) have no impact on the new environment
(B) have greater impact on the new environment
(C) have limited impact on the new environment
(D) are seldom transported to another environment
(E) don't survive in a new environment

4. When DNA is combined in ways to augment certain characteristics in a species' natural growth, the result is known as a

(A) nonnative species
(B) relict species
(C) transplanted organism
(D) genetically modified organism
(E) nonviable organism

5. When nontarget fish species or other marine animals are caught in nets or through other fishing methods, they are called

(A) surplus species
(B) bycatch
(C) alternate species
(D) trash fish
(E) game fish

6. The huge increase in yields obtained through modern cultivation techniques has become known as the

(A) GMOs
(B) cross-pollination methods
(C) aquaculture dilemma
(D) genetic revolution
(E) green revolution

7. Small land plots that are planted early, heavily sprayed with pesticides, and their crop later destroyed are known as

(A) hectares
(B) sustainable agriculture
(C) trap plots
(D) testers
(E) polluted plots

8. Which of the following has caused the greatest increase in the introduction of marine alien species into new ecosystems?

(A) Tourism
(B) Air travel
(C) Shipping
(D) Commercial fishing
(E) Home aquariums

9. Fishing methods like bottom trawling, gill netting, and longlining are especially subject to

(A) crew safety policy
(B) seasonal variation
(C) high fuel costs
(D) poor yields
(E) bycatch

10. The Food Quality Protection Act caused the Environmental Protection Agency to

(A) assess inert ingredients for the first time
(B) raise levels for pesticides
(C) determine whether carcinogens existed in food
(D) protect the elderly from DDT
(E) send hybrid rice to third-world countries

11. Agroecology is another word for

(A) tillage
(B) sustainable agriculture
(C) degenerative farming
(D) strip mining
(E) historical agriculture

12. Something that lowers the quantity, quality, or value of resources is known as a

(A) growth factor
(B) decimator
(C) biocide
(D) agricultural sink
(E) biological pest

13. When species are transported geographically to an unknown area, they are called

(A) native species
(B) game fish
(C) exotics
(D) nonnative species
(E) functional species

14. Three agencies are responsible for regulating pesticides used in food: the Environmental Protection Agency, the Food and Drug Administration, and the

(A) Centers for Disease Control
(B) Department of Health and Human Services
(C) Department of Agriculture
(D) National Institutes of Health
(E) Department of Homeland Defense

15. Rodenticides kill rats and mice, while ticks and spider mites are killed by

(A) aracnicides
(B) mitocides
(C) fungicides
(D) acaricides
(E) herbicides

16. It has been estimated that over 60% of all processed food in America contains

(A) GMO ingredients
(B) fatty acids
(C) carcinogens
(D) DDT
(E) phosphorus

17. Hormones that stop larval growth, combined with mechanical vacuuming of bugs from plants, are examples of

(A) crop rotation
(B) low input agriculture
(C) integrated pest management
(D) cross discipline activation
(E) plant desertification

18. DDT, which kills malaria-carrying mosquitoes, had a large negative impact on

(A) horses
(B) local ladybugs
(C) earthworms
(D) nontarget predatory bird species
(E) migrating robins

19. One method to conserve water and make sure it gets directly to plant roots is through

(A) sieve sleeve irrigation
(B) drip irrigation
(C) rolling sprinklers
(D) flood irrigation
(E) crop circle sprinklers

20. Organic farms have lower feed and veterinarian costs since their animals are

(A) smaller in size
(B) crowded into barns
(C) healthier
(D) raised in larger herds
(E) fed GMO ingredients

› Answers and Explanations

1. C—Besides changing native marine ecosystems, alien species can also affect human health.

2. E—Aquaculture may reduce the pressure on native species as food sources.

3. B—Unlike other oceans, there is little natural exchange between these oceans.

4. D—Altering genetics makes some people worry about the potential for abuse.

5. B—Dolphins were a bycatch of tuna fishing until new escape nets were mandated.

6. E

7. C—Insects in the area swarm to the early crop and are killed, lowering the need for pesticides later.

8. C—Ships using ocean water for ballast, flush it in port along with acquired species.

9. E—Many different species are caught by their gills, for example.

10. A—The Act sets limits on combined pesticide exposures and their safety.

11. B

12. E

13. D—Nonnative species are also known as alien species.

14. C

15. D

16. A—Genetically modified organisms or genetically enhanced growth.

17. C—Integrated pest management uses a combination of methods.

18. D—DDT caused birds' eggshells to be very thin and break when the mother nested.

19. B—Watering is close to the plants instead of being sprayed in the air all over the field.

20. C—Since they aren't stressed or in tight pens, they don't pass on disease as easily.

Free-Response Questions

1. Humans have wanted to control pests for thousands of years. By trial and error they found different materials and methods that protected themselves, their crops, and their animals from pests. Modern pest control is performed with pesticides (chemicals that kill or repel certain pests). It is important to consider the end consumer of agriculture and aquaculture products. Choosing cultivation methods carefully is important.

   (a) Explain three disadvantages of pesticide use.
   (b) How does polyculture farming help alleviate some of the problems associated with aquaculture?
   (c) What are the three main crops eaten by humans? Are they the same for people all over the world?

Free-Response Answers and Explanations

1.

a. There are numerous disadvantages to using pesticides. As seen with DDT, pesticides can find their way into a food chain, poisoning many species not directly related to or harmful for the crop being protected. Pesticides also create genetically stronger species, forcing us to come up with stronger and more dangerous pesticides to combat pests. Pesticides also directly harm the environment by seeping into soils and groundwater over time, poisoning a large section of land and associated groundwater.

b. By establishing different species in the same area, a polyculture mimics natural environments and is more environmentally sound than a monoculture, uses less food, and produces more.

c. The three main grain crops eaten by humans are wheat, rice, and maize. Potatoes, barley, rye, and oats, grown at higher altitudes are important in northern Europe and Asia. Roots and tubers are grown in greater numbers in equatorial climates.

❯ Rapid Review

- Nonnative (alien) species are transplanted geographically to a formerly unknown area.
- Accidental catching and killing of nontarget species, like nonmarketable "trash" fish and marine mammals (dolphins, sharks), have impacted ocean ecosystems.
- Malnourishment is a nutritional imbalance caused by the lack of important dietary elements or the inability to absorb essential nutrients from food.
- The most common nutritional problems are anemia (low iron in the blood), goiter (too little iodine for mental development, metabolism, and thyroid function), and vitamin A deficiency (causes blindness).
- Obesity is defined as more than 30 pounds over a person's ideal weight for their height and age.
- The three main grain crops eaten by humans are wheat, rice, and maize (called corn in the United States).
- Although thousands of species of plants have been grown for food over time, only 16 species are currently cultivated.
- Genetic engineering is the bioprocess of taking DNA (blueprint proteins) from one species and splicing them into the genetic material of another.
- When DNA is combined in ways to augment certain characteristics in a species' natural growth, the result is known as a genetically modified organism (GMO).
- Drip irrigation provides water directly to thirsty plant roots.
- A biological pest is something that lowers the quantity, quality, or value of resources.
- Insects make up roughly 75% of all species on the Earth.
- Pesticides are chemicals that kill or repel certain pests, while biocides kill a broader range of organisms.
- Chemicals that kill specific targets are named accordingly, such as insecticides (kill insects), herbicides (kill plants and weeds), acaricides (kill ticks and spider mites), aracnicides (kill spiders), rodenticides (kill rats and mice), and fungicides (kill molds and fungus).
- Pesticide buildup in the top members of the food chain is known as biomagnification.
- Trap crops are small plots planted before the rest of the field and heavily sprayed with pesticides to trap and kill insects that would harm the main crop.
- In sustainable agriculture (regenerative farming and agroecology), farmers grow nutritious crops while turning away from harmful past practices.
- When nontarget fish species or other marine animals are caught in nets or through other fishing methods, they are called bycatch.

Forestry and Rangelands

IN THIS CHAPTER

Summary: One of the Earth's wonders is its diversity of plant life, including ancient forests. Although forests are complex ecosystems in themselves, they are also divided according to climate. Deforestation and poor land practices often lead to overgrazing and desertification.

Keywords

✪ Old-growth forest, deforestation, overgrazing, desertification, extinct, relict species, degradation, tree plantations, forest fires, forest/rangeland management

---

Forests

Forests and woodlands cover nearly 30% of the global land surface. Most of this area is known as *closed canopy* (tree crowns cover over 20% of the ground below). The rest is *open canopy* with less than 20% ground coverage.

Forests are mainly divided into *temperate* (moderate climate) or *tropical* regions. Temperate forests are grouped into *conifers* (needle-leaf trees) like pine, spruce, redwood, cedar, fir, sequoia, and hemlock, while tropical forests contain flat-leaf trees. Old-growth forests contain mostly conifers. Second- and third-growth forests contain trees of the same age and size as some of the younger old-growth forests, but have far fewer plant and animal species.

TIP

Temperature and rainfall are the major determinants of forest type, including temperate rain forest, tropical dry forest, and tropical rain forest. A *temperate rain forest* is found in only a few special places around the world, such as the Pacific temperate rain forest on the west coast of North America. These temperate forests are often dominated by conifer trees adapted to wet climates and cool temperatures.

Located near the equator, a *tropical dry forest* has distinct rainy and dry seasons. Most tropical dry forest plants have adapted to withstand high temperatures and seasonal droughts.

The third major forest type receives and contains a lot of moisture. A *tropical rain forest,* also found near the equator, harbors the richest diversity of terrestrial plant and animal species. Though hot and humid, these forests often get less than half the rainfall of a "rain" forest and have a more open canopy with less dense undergrowth.

Today, the largest and most severe species loss is taking place in the tropical rain forests near the equator. Trees are cleared for grazing, farming, timber, and fuel. Ecologists have found that over 50% of the the Earth's original rainforests, since prehistoric times, have been cleared. At current rates, the remaining forests are being lost by around 1.8% per year.

Old-Growth Forests

Old-growth forests are found primarily in northern climates, although there are small caches of untouched trees in remote locations like Tasmania.

Old-growth forests are made up of trees that are often several thousand years old and take at least 100 years to regrow to maturity if cleared.

Old-growth forests are those that have never been harvested. They contain a variety of trees between 200 and 2,000 years old. Forest floor leaf litter and fallen logs provide habitat for a complex interdependent mix of animals, birds, amphibians, insects, bacteria, and fungi, which have adapted to each other over geological time. When forests are cleared, the plants, animals, birds, and insects living under their protective cover are displaced or destroyed as well. Genetic uniqueness (biodiversity) of these affected species is permanently lost.

Redwood National Park and three California state parks contain some of the world's tallest trees: old-growth, coastal redwoods. Living to between 2,000-2,500 years old, they grow to be over 110 meters (370 feet) tall. Other species such as spruce, hemlock, Douglas fir, and sword ferns create a multilevel tree canopy towering above the forest floor.

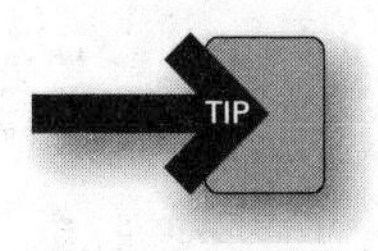

Redwoods are long lived because they are particularly resistant to insects and fire. Scientists discovered giant redwoods contain high levels of bark tannins protecting them from disease. Additionally, redwoods can grow from seeds or new sprouts from a fallen tree's root system, stumps, or burls. So clearing of fallen trees reduces new growth.

The most important factor in redwood survival is their biodiversity. Forest floor soils play a big role in tree growth. A healthy redwood forest includes a variety of tree species, as well as ferns, mosses, and mushrooms. These are important to soil regeneration. Fog from the nearby Pacific provides cooling and moisture for the trees.

The Redwoods, however, have lost a lot of ground. Of the original 1,950,000 acres of redwood forests growing in Oregon and California, only 5% of the original old-growth coast redwood forest remains. Three percent of these acres are preserved in public lands and 1% is privately owned and managed.

It comes as no surprise that when trees are reduced through overcutting, biodiversity drops. When forests or grasslands are cleared and planted as a single cash or food crop, the number of species drops to one, plus a few weeds. But this is only part of the problem. Since forests support animal species with food and shelter, these species are also eliminated. Often, new species replace the originals, but generally the total number of species declines. When plant cover is removed, other neighboring populations (mammals, birds, and insects) are greatly affected.

If a species can't find new habitat or adapt to changed land use, it often becomes *extinct.* Its unique genetic information and position in the ecosystem are forever lost.

A species that has survived, while other similar ones have gone extinct, is called a *relict species.* A relict species, like the European white elm tree in western Siberia, may have had a wider range originally but is now found only in specific areas. Other relict species, like horseshoe crabs or cockroaches, have survived unchanged since prehistoric times, even as other species become extinct.

Deforestation

From the time when early peoples switched from being hunters and gatherers to settling down and growing crops, humans have had more and more impact upon the land. The clearing of trees to grow food crops is so common that there are few original forested areas still untouched.

Farmers in the infant United States cleared the land and put it to the plow. The timber was used to build homes, farm buildings, and fences. Cleared areas around a wooded farm provided safety from forest predators such as bears, wolves, and cougars.

Gradually, landowners realized some methods of timber harvesting were far better than others. This became noticeable when cleared land, once rich in nutrients, grew poorer and poorer for raising crops and began to erode in locations continually exposed to wind and water.

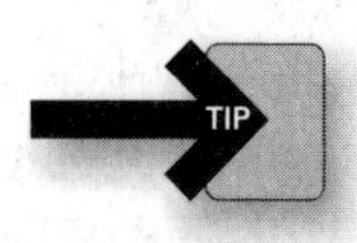

Fuel wood and charcoal are another major human requirement, especially in developing countries. It has been estimated that over 1.5 billion people currently depend on fuel wood as their main fuel source. By 2025, as the global population increases, fuel wood demand is estimated to become twice the available supply.

Fortunately, Asian deforestation has slowed from 8% in the 1980s to less than 1% today. The number of protected areas has also increased over the past 20 years, from around 2.6 million km^2 to about 12.2 million km^2.

KEY IDEA

Deforestation is the large-scale destruction of trees in an area by disease, cutting, burning, flooding, erosion, pollution, or volcanic activity.

Large lumber companies must be economically successful or they go out of business. They have to get as much timber from the land as possible to meet growing needs for wood, paper, and other products. As with all industries, there are good players and bad players. Good players work with scientists to plant and maintain this renewable resource. Others grab what they can and don't consider the long term.

When a forested area is cut down completely without a tree left standing, it is known as *clear-cutting.* In the United States and Canada, lumber and pulpwood is most often harvested using this method. Clear-cutting is more economical than selective cutting since it allows the use of large machines to fell, trim, and truck logs much faster. However, the major drawbacks to clearing a wooded area are that it completely wastes small trees, invites erosion, and destroys habitat for other species.

Clear-cutting performed in small scattered areas, individual rows, or alternating strips has less overall impact and more potential for faster regeneration than huge deforested areas encompassing thousands of acres. As in real estate, the location and geography of a clear-cut area are important. Clear-cutting a steep slope will cause much more erosion than clear-cutting a flat area. In addition to stripping the land of soil and nutrients, clear-cutting allows soil runoff to fill streams and kill aquatic life.

In *selective cutting,* only a small percentage of mature trees are cut every 10 to 20 years. Ponderosa pines, for example, are selectively thinned to improve the growth of remaining trees.

Leaving mature trees standing keeps them as a seed source for new tree growth in a cleared area and provides forest habitat.

Industrial timber is used for lumber, plywood, veneer, particleboard, and chipboard. These products make up about half of the world's use of timber. International timber trade amounts to over $100 billion per year, with developed countries producing less than half the timber total, but using about 80% of the available timber. Canada, Russia, and the United States are the largest producers of industrial wood and paper pulp, but much of North American logging is done in managed forests where downed trees are replaced with new seedlings.

Forest Management

Of the world's forests, around 25% is managed for wood output. Sustainable harvests are key, and special attention is paid to insect, disease, and fire impacts.

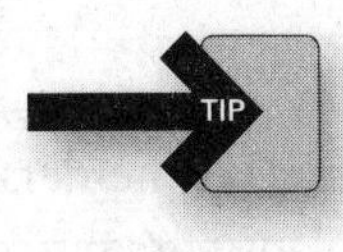

It has been estimated that 50% of all the Earth's forests have been converted to cropland depending on the species. Although most countries cut down more trees than they replant, China and Japan have developed huge reforestation programs. Japan has increased its forests to nearly 68% of its land area, and imports the majority of its wood needs from its trading partners.

Government forest management is controversial since lumber companies, ecologists, landowners, developers, and fishermen have different opinions on how forest areas should be used and what access should be allowed. Lumber and jobs come into the equation, as well as public demand for forests to offer wilderness hiking, camping, and wildlife habitat. Policy changes have opened public lands to more logging, mining, oil and gas drilling, and motor vehicles.

Tree Plantations

Reforestation projects involve large plantings of a single species like eucalyptus or pines. Although one-species planting makes it easier for mechanical harvesting, it also invites insect and disease devastation. Monoculture tree planting decreases the strength found in a diverse natural or planted forest. A species-specific disease isn't able to wipe out an entire area planted with different kinds of trees.

Tree planting is not just a commercial trend. Communities are planting vacant lots and roadways with fruit and nut trees, as well as fast-growing trees that enrich the soil, curb erosion, and serve as wind breaks.

Forest Fires

In the 1930s, hundreds of millions of hectares were destroyed by fire that leveled entire towns and killed hundreds of people. Public outcry demanded the U.S. Forestry Service do something. So a policy of extinguishing fires as quickly as possible was enacted. An advertising campaign featuring Smokey the Bear warned that "only *you* can prevent forest fires." People were educated about carelessness with campfires and not to toss lit cigarettes from cars.

Unfortunately, only part of the picture was addressed. Some biological species need fire to reproduce because a forest fire clears undergrowth allowing new sprouts to survive. Also, when a lot of forest debris gathers, it increases rather than decreases the chance of large, more devastating wildfires.

In 2012, the federal government spent $2.8 billion to fight over 75,000 wildfires. Controlled burns and preburns, besides making forests less fire-prone and removing dead wood and debris, cost taxpayers only about 50 cents per hectare. The key is to monitor natural fires and use scientific methods of control to gain their benefits.

Rangelands

Trees are important for many reasons, but grasslands are another big biome. Composed of savannas, steppes, prairies, open woodlands, and a variety of grasslands, rangelands cover around 25% of the planet's land surface. Much of North America's Great Plains (e.g., 3.8 billion hectares) is grassland. Another 4 billion hectares of other lands (i.e., tundra, marsh, scrub brush, forest, and desert) support approximately 3 billion cattle, goats, sheep, buffalo, and camels that humans use for nutrition. The key is to balance sustainable herding and grassland biodiversity.

Unfortunately, grassland biomes are often converted to cropland or urban areas. In fact, annual grassland conversion is three times that of tropical forests worldwide. The U.S. Department of Agriculture cites more threatened plant species in rangeland than in any other biome.

Rangeland Management

Careful monitoring and moving of grazing animals by ranchers and *pastoralists* (i.e., herders) plays a big role in rangeland management. Rainfall, alternate and seasonal plantings, varied forage, and natural fertilization methods all help maintain range quality.

Rangelands in the United States are often overgrazed due to political and economic pressures. The Natural Resources Defense Council reports that 30% of public rangelands are in fair condition, while 55% are in poor to very poor condition. Although regulations exist to protect rangeland, poor enforcement and limited funds augment overgrazing. Decreasing native forage, invasive nonnative species, and higher erosion are added factors.

Grazing

Range specialists have found that brief (a day or two), intense, *rotational grazing* like that done by wild herds (e.g., buffalo, zebras) is often better for the land than long-term grazing. A herd confined to a small area eats everything, not just the tender shoots, tramples woody plants and weeds, and fertilizes heavily before moving to greener pastures. The herd's impact and subsequent fertilizer application provide rangeland with needed nutrients, while competing weeds are eliminated.

Raising wild species (e.g., impala, red deer) also supports grazing sustainability. Wild species are often more pest resistant, disease resistant, and drought tolerant, and can fend off predators better than domestic cattle, goats, or sheep. In the United States, modern ranchers note that elk, American bison, and various African species need less care and feeding than cattle or sheep and bring a better market price.

Deserts

Most people think of *deserts* as uninhabitable and devoid of life. These barren stretches are probably the last place on Earth most people would choose to visit or live. Baking daytime temperatures up to 57°C, and cold nights, down to 5–10°C, take people by surprise. Extreme heat and dryness, interrupted by unpredictable flash floods, give places like Death Valley and The Desert of No Return (the Taklimakan Desert in northwest China) their names. In fact, the Antara desert in Chile has spots that only receive rain a few times a century, and some areas have never had rain.

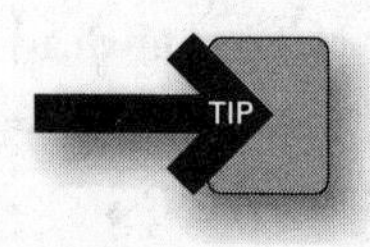

Deserts cover roughly 30% of the Earth's surface. Most deserts, like the Sahara in North Africa and the deserts of the southwestern United States, Mexico, and Australia, are found at low latitudes. Cold deserts are mostly found in the basin and range area of

Table 13.1 The world's major deserts cover a lot of area around the globe.

DESERT	LOCATION	APPROXIMATE AREA (km^2)
Sahara	North Africa	9,065,000
Arabia	Saudi Arabia	2,240,000
Gobi	China	1,295,000
Kalahari	Southern Africa	582,000
Chihuahuan	North central Mexico, southwestern United States (Arizona, New Mexico, Texas)	455,000
Great Basin	United States (Idaho, Oregon, Nevada, Utah)	411,000
Great Victoria	Australia	338,500
Patagonia	Peru, Chile	150,000

Utah and Nevada and in sections of western Asia. Table 13.1 lists the locations and areas of several of the world's largest deserts.

The common misconception that the desert is a lifeless landscape is really an illusion. Deserts have abundant and unique plant life, as well as unique animals. Soils often contain many nutrients just waiting for water, so that plants can spring to life when it rains. Fires provide a carbon source (ash), and flash flooding mixes everything up to mineralize desert soils.

The dual problems of storing water and finding shelter from the sun's blistering heat and the night's cold make deserts an environment with few large animals. Snakes and lizards are dominant creatures of warm deserts. Only small mammals, like the kangaroo rat of North America, are found in any great number.

Types of Deserts

In 1953, Peveril Meigs divided the Earth's desert regions into three categories (*extremely arid, arid,* and *semiarid*) according to their annual rainfall. In Meigs' system, extremely dry lands can go over a year without rain. Arid lands have less than $\frac{1}{4}$ cm of rainfall yearly, and semiarid lands have a mean rainfall of between $\frac{1}{4}$ and $\frac{1}{2}$ cm. Arid and extremely arid lands are called deserts, while semiarid grasslands are known as *steppes.*

While deserts can be classified in different ways, most deserts are distinguished by the following factors:

- Total rainfall
- Number of days of rainfall
- Temperature
- Humidity
- Location
- Wind

Deserts are not limited by latitude, longitude, or elevation. They are found everywhere from the poles to the equator. The People's Republic of China has the highest desert, the Qaidam Depression (2,600 meters above sea level), and one of the lowest deserts, the Turpan Depression (150 meters below sea level).

Desertification

Deserts are formed from a variety of geological and natural factors. Global temperatures, rainfall rates, and tectonic processes all contribute. *Desertification* is the downgrading of rich soil and land into dry, barren lands. Today desertification takes place because of damaging human activities and climatic changes.

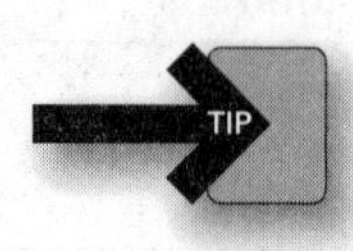

Desertification occurs because dry land environments are vulnerable to overdevelopment and poor land use. The following factors all increase desertification:

- Poverty
- Political instability
- Deforestation
- Overgrazing
- Rain runoff
- Changing wind patterns driving away rain clouds
- Bad irrigation practices

Estimated calculations indicate over 250 million people worldwide are directly harmed by desertification. Moreover, it is thought that as much as one billion people in nearly 100 countries are at risk. These populations are poor and have few resources to stem the desertification problem.

Degradation of Dry Lands

When formerly fertile land is degraded, there is a loss of biodiversity. As a result, the economic yield and complexity of cropland, pasture, and woodland is damaged. *Degradation* is due mainly to climate changeability and unsound human activities, like overgrazing and deforestation. While drought is often linked with land degradation, it is a natural event. It happens when there is much less than normal rainfall in a region over an extended period of time.

Rainfall

Deserts have little if any freshwater supplies. In desertification, rainfall varies seasonally during the year, with wide swings happening over years and decades, leading to drought. Over time, desert ecology adjusts to this moisture fluctuation.

The biological and economic resources of dry lands, like soil quality, water supplies, plants, and crops, are easily damaged. For centuries, native people have protected these resources with time-tested methods like *crop rotation* and nomadic herding.

Unfortunately, in recent decades rising populations and deteriorating economic and political conditions have contributed to poor land practices. Also, when people don't take climate and soil conditions into account and respond accordingly, desertification is the result.

Land Overuse

Land overuse can come from economic circumstances, poor land laws, or cultural customs. Some people exploit land resources for their own gain with little thought for the land or neighboring areas. Some people in poverty have little choice but to overuse the meager resources available to them, even to the extent of wearing out the land.

Trade and exploitation of a country's natural resources often leaves land restoration in the hands of local people with little or no funding. In the same way, an economy based on the sale of crops can force farmers to ignore the land's overexploitation.

Refugees and other displaced people damage rich land during wars and national emergencies. Natural disasters like floods and droughts do the same thing.

Population Density

Most people blame desertification on overpopulation. However, large populations practicing good conservation and land management can avoid desertification. It is said the United States could feed most of the world with grain produced and stored in the midwestern grain belt.

Desertification is a complex problem, and the relationship between population and desertification is not clear. Sometimes population decreases cause desertification, since there are fewer people to take care of the land. In some countries when young villagers go to the city to find work, their aging parents can't keep up with the cultivation needs of the land.

Soil and Vegetation

When topsoil is blown away by the wind or washed away by rainfall, the physical structure and biochemistry of the land is changed. Surface cracks appear and nutrients are lost. If the water table rises from poor drainage and irrigation methods, the soil is saturated and salts increase. When cattle trample soil, it's harder for plants to grow in the compressed soil and the soil erodes.

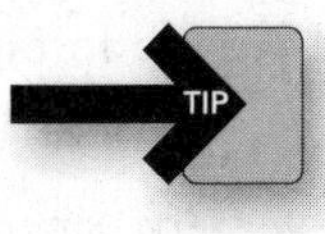

Less plant growth is both a result and cause of desertification. Loose soil damages plants, buries them, or exposes their roots. Disturbed land also increases downstream flooding, poor water quality, river and lake sedimentation, and silt buildup in reservoirs and navigation channels. It can cause dust storms and air pollution, damaging machinery and reducing visibility. Wind-blown dust increases public health problems, including allergies, eye infections, and upper respiratory problems.

Economics

More and more people and governments are seeing the link between desertification, displaced people, and military conflicts. In the past 30 years, many African people have relocated or been forced to move to other countries because of war, drought, and land degradation. Environmental resources in and around cities and refugee camps, where these people settled, came under severe pressure. Difficult living conditions and the loss of cultural identity further undermined social stability.

Little economic data on losses resulting from desertification exist, but it is thought the depletion of natural resources in some countries is as high as 20% of their annual gross domestic product (GDP).

Trends

Soil loss, low moisture, and high temperatures are all factors in desertification studies. Researchers in Brazil, Namibia, South Africa, the southwestern United States, and northern Mexico are all actively examining land degradation and sand movement. Large global deserts are also monitored for geological mineral mapping and the determination of the active sand transport corridors.

Remote Sensing

In the past 70 years, the use of remote sensing has been an important tool in the study of dynamic desert features, like dunes. The ability to study changes over time allows analysis of current climatic systems and marginal land areas at risk for future desertification.

Remote sensing helps geologists inspect an entire desert system over different time periods. It provides a way to project past activity and monitor current climate changes. Drought-prone areas on the margins of wind pathways and sand seas are prone to dune advance and encroaching desertification. Human activity in these regions must also be examined for climatic impacts.

› Review Questions

Multiple-Choice Questions

1. Deserts cover roughly what percentage of the the Earth's land surface?
   (A) 10%
   (B) 20%
   (C) 30%
   (D) 40%
   (E) 50%

2. When formerly fertile land is degraded, there is
   (A) loss of biodiversity
   (B) an increase in water
   (C) a gain in economic yield
   (D) species diversification
   (E) a drop in fungal growth

3. Deforestation, overgrazing, and bad irrigation practices all contribute to
   (A) juvenile delinquency
   (B) glacier formation
   (C) desertification
   (D) wetlands formation
   (E) sedimentation

4. In size, the Great Basin Desert of the United States is only about one-third the size of the (see Table 13.1)
   (A) Kalahari Desert
   (B) Gobi Desert
   (C) Chihuahuan
   (D) Sahara
   (E) Patagonia

5. Forests are mainly divided into temperate and
   (A) dry regions
   (B) wetland regions
   (C) arctic regions
   (D) tropical regions
   (E) desert regions

6. Old-growth coastal redwoods live to be as much as
   (A) 300 years old
   (B) 750 years old
   (C) 1,000 years old
   (D) 1,500 years old
   (E) 2,000 years old

7. The following factors can all increase desertification except
   (A) wind
   (B) total rainfall
   (C) forest management
   (D) temperature
   (E) location

8. Second- and third-growth forests contain
   (A) many more plant and animal species
   (B) trees of the same age and size
   (C) mainly conifers
   (D) larger trees than those in old-growth forests
   (E) only broad-leafed trees

9. The European white elm tree in western Siberia is known as a
   (A) new species
   (B) nonnative species
   (C) relict species
   (D) primary species
   (E) replanted species

10. Which is important for some biological species to reproduce and clear undergrowth for new sprouts?
    (A) Lime
    (B) Fire
    (C) Iron oxide
    (D) Ice
    (E) Wind

11. Remote sensing is an important tool in the study of dynamic desert features like
    (A) tree growth
    (B) camel populations
    (C) flash floods
    (D) dunes
    (E) locust species

12. A tropical rain forest, found near the equator, has
    (A) primarily conifers
    (B) few native species
    (C) increasing fungal growth
    (D) the richest diversity of terrestrial plants and animals
    (E) no new species to be observed by scientists

13. In developing countries, the need for which major resource will double in the next 25 years?

(A) Fuel wood and charcoal
(B) Soybeans
(C) Milk
(D) Television
(E) Latex

14. What percentage of all the forests on the Earth has been converted to cropland?

(A) 10%
(B) 20%
(C) 30%
(D) 40%
(E) 50%

15. Deserts are distinguished by all the following factors except

(A) number of days of rainfall
(B) wind
(C) temperature
(D) oases
(E) lightning strikes

16. When plant cover is removed by clear-cutting, other local populations (mammals, birds, and insects)

(A) increase
(B) are not affected
(C) are greatly affected
(D) remain in the area
(E) resist disease

17. When topsoil is blown or washed away, what happens to the remaining land?

(A) Its biochemical makeup is changed.
(B) Its nutrients are maintained.
(C) Its structure is unchanged.
(D) It becomes a major conservation factor.
(E) Its water content increases.

18. What extremely long-lived tree species is resistant to insects and fire?

(A) Hickory
(B) Spruce
(C) Redwood
(D) Aspen
(E) Oak

19. Rotational grazing involves all the following ranching and herding activities except

(A) confining herds to a small area
(B) herds fertilizing heavily
(C) herds trampling woody plants and weeds
(D) long-term grazing of herds over a large area
(E) herds eating everything in the area

› Answers and Explanations

1. C

2. A—As species are killed or forced to move, the diversity of an area declines.

3. C—The soil loses its nutrients and is less likely to support many species.

4. B

5. D—Tropical forests grow well in hot and humid equatorial regions.

6. E—Scientists have found them to contain compounds resistant to insects.

7. C—Forest management decreases desertification.

8. B—These have far fewer plant and animal species than do old-growth forests.

9. C—Relict species outlive other species but often lose much original habitat.

10. B—Fire contributes organic mass to soil and exposes the ground to more light.

11. D—Remote sensing is needed for the "big picture" of dune growth and travel.

12. D—High rainfall and warm temperatures allow many different species to thrive.

13. A—Fuel wood and charcoal are used for basics such as cooking and heating.

14. E

15. E—Since they have few storms and rain, they have few lightning strikes.

16. C—Clear-cutting destroys not only forests but also their dependent species' habitat.

17. A—Surface cracks appear, nutrients are lost, and excess water may increase salts.

18. C

19. D—Rotational grazing involves short-term grazing over a small area.

Free-Response Questions

1. Wildlife conservationists want a ban on cattle grazing to stop the threat to endangered species in the southwestern United States. Although grazing fees are charged to ranchers for using federal lands, they are quite low and serve as a subsidy to ranchers who pay much more to lease private land. Over 20,000 permits/leases are issued annually according to the U.S. Bureau of Land Management for grazing on 157 million acres of federal rangelands. These permits/leases bring in nearly $12 million annually. However, the administration and maintenance of those same lands cost taxpayers roughly $50 million per year. This fiscal discrepancy serves as a wake-up call to rangeland management and public policy makers. Defending western culture and history, ranchers also avoid subdivision of the West by developers, along with the resulting environment and wildlife impacts.

(a) In what ways does grazing benefit grassland biomes?
(b) Describe major environmental and socioeconomic consequences of overgrazing.
(c) Define desertification.
(d) Name five causes of desertification.

2. It has been estimated that over 50% of all the forests on the Earth have been converted to cropland. Most countries cut down more trees than they replant, but some are starting huge reforestation programs.

(a) Explain why reforestation often increases damage from forest diseases.
(b) Explain why governmental management of forests is controversial.

Free-Response Answers and Explanations

1.

a. Managed properly, grazing can increase wildlife diversity and the overall health of grasslands. When herds are allowed to graze rotationally for short amounts of time, they eat everything, not just the tender shoots; trample woody plants and weeds; and fertilize heavily before moving to greener pastures. This keeps the land healthy and supports native forage species. It also promotes areas of varied growth height, which support a variety of habitats and niches for diverse wildlife to fill.

b. Overgrazing is one of the major reasons for desertification. By reducing the land's ability to support life, desertification negatively affects wild and domestic animals, plant life, and people. Because it causes soil erosion while also affecting the ability of the soil to soak up water, the water loss drives away or kills local flora and fauna. Socioeconomic consequences for humans also result from desertification brought on by overgrazing. Whole human populations must move from land unable to sustain their animals and crops or provide water. The rippling effects from overgrazing can be seen from habitat destruction to war since people will fight for fertile land and crop resources.

c. Desertification is the downgrading of rich soil and land into dry barren lands.

d. Deforestation, overgrazing, rain runoff, changing wind patterns driving away rain clouds, and bad irrigation practices.

2.

a. Many reforestation plantings are done with a single tree species. When this happens, the forest becomes much more vulnerable to disease than when it had a wide variety of tree species resistant to different diseases.

b. Governments must juggle the different needs of lumber companies, ecologists, landowners, developers, fishermen, and the public.

❯ Rapid Review

- When a forested area is cut down completely, it is known as clear-cutting.
- Old-growth forests are those that have never been harvested, with trees between 200 and 2,000 years old.
- Forest floor leaf litter and fallen logs provide habitat for interdependent animals, birds, amphibians, insects, bacteria, and fungi adapted to each other over geological time.
- When the forests are cleared, plants, animals, birds, and insects are displaced or destroyed, causing a loss of biodiversity and often extinction.
- Forests are mainly divided into temperate (moderate climate) or tropical regions.
- Temperate forests are grouped into conifers (needle-leaf trees) like pine, spruce, redwood, cedar, fir, sequoia, and hemlock, while tropical forests contain flat-leaf trees.
- Old-growth forests contain mostly conifers.
- Second- and third-growth forests contain trees of the same age and size as some of the younger old-growth trees, but have far fewer plant and animal species.
- Land overuse can come from economic circumstances, poor land laws, or cultural customs.
- Some biological species need fire to clear undergrowth, allowing these species to reproduce, and new sprouts to survive.
- When topsoil is blown or washed away, the remaining land's physical structure and biochemical makeup are changed.

- Population decrease can cause desertification, since there are fewer people to take care of the land.
- Desertification is the downgrading of rich soil and land into dry, barren lands.
- An economy based on crop sales can cause farmers to ignore the overexploitation of the land.
- A temperate rain forest is found in only a few special places around the world, such as the Pacific temperate rain forest on the west coast of North America.
- Intense, rotational grazing like that done by wild herds (e.g., buffalo, zebras) is often better for the land than long-term grazing.
- Wild species are often more pest resistant, disease resistant, and drought tolerant, and can fend off predators better than domestic cattle, goats, or sheep.
- Elk, American bison, and various African species need less care and feeding than cattle or sheep, while bringing a better market price.

Land Use

IN THIS CHAPTER

Summary: Population increases have changed land use from rural to urban. Development and suburban living is much more common. Private and public resource management is under debate. Sustainability of forests, wildlife refuges, national parks, and wilderness areas must be addressed. Mining restoration and recycling are important for clean air, water, and long-term environmental stewardship.

Keywords

✪ Urbanization, megacity, suburban sprawl, ecological footprint, preservation, remediation, mitigation, restoration, strip mining, recycling

Land Use

Since ancient times, people have lived in rural areas where resources came from fishing, hunting, farming, herding, and mining. Occasionally, they gathered to sell, trade, and buy necessities. Over time, some found jobs and stayed in the gathering places permanently. This gave rise to villages and groups of homes joined by family ties, culture, tradition, and the land.

Historically, towns and cities grew into urban areas with large populations specializing in arts, crafts, services, or professions not necessarily tied to local resources and land. Cities expanded out from seaports and major rivers. Imported and exported resources, along with industrialization, brought prosperity to inhabitants, increased jobs, and drew even more people and industries to the city. This gradual change in land use and culture is known as *urbanization.*

Over 80% of the U.S. population lives in cities. Currently around 50% of the world's population lives in cities, and that number is projected to be over 60% by 2035. In fact, China plans to build 400 new cities of $\frac{1}{2}$ million residents each in the next 20 years.

One hundred years ago, London was the only city with over 5 million people. The United Nations Population Division now estimates that 30 cities have larger populations than that. Today, many cities have become *megacities* (i.e., with over 10 million inhabitants). It is projected that by 2015, a dozen cities will have populations between 15 and 30 million residents, 75% of these in the developing world. Additionally, millions of undocumented, temporary workers work in urban areas (e.g., Beijing) daily or seasonally.

Urban Growth Factors

Urban populations increase naturally by more births than deaths and by immigration. Natural increases come from good and plentiful food and water, sanitation, and medical services. Globally, natural increases account for the majority of population jumps (e.g., Eastern Asia and South America). In Western Asia and Africa, immigration to cities for employment and services is a leading cause of urban growth.

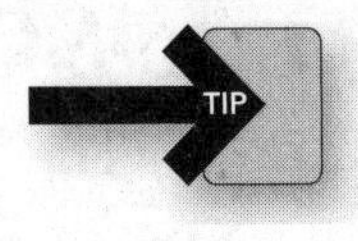

Immigration is affected by different characteristics or *push and pull factors.* Overpopulation, conflicts, and lack of resources in a country push new generations into the cities for work and a better life. Communications, arts, and an exciting, modern lifestyle pull people into urban areas as well.

Urban Problems

In large cities, resources and land are strained if not exhausted. For example, in Atlanta, Georgia, and Phoenix, Arizona, strict summer water rationing is mandatory since public water resources are shrinking and nearly exhausted during drought periods.

Traffic congestion, fuel consumption, and subsequent air pollution is a huge problem in densely populated cities such as Jakarta, Indonesia. The average worker spends between three and four hours traveling to and from work each way from suburban areas due to traffic jams. Underdeveloped countries, with the added burden of belching factories, coal and wood fires, and vehicle exhaust, have an ever-present gray veil of pollution affecting their citizens. Lung cancer deaths from smoking and pollution in Shanghai, China, are five to seven times higher than in the countryside.

Developing countries struggle with sewage treatment and proper sanitation. With no money to build modern facilities, only a small percentage of urban residents have access to adequate sanitation. In Latin America, only 2% of urban sewage is treated. In Cairo, Egypt, aging treatment facilities designed to provide for around 2 million inhabitants are servicing over 10 million people.

In Mexico City, roughly 50% of the population (20 million) lives in shantytowns built on undeveloped land from scavenged materials such as packing crates, plastic, and brush. There is no clean water or sanitation. Sometimes people build on hazardous waste sites purposely left undeveloped. When the government removes the shacks from these hazardous areas, people either rebuild or move to another unauthorized area.

Suburban Sprawl

People in developed countries, disenchanted with urban crush, crime, and congestion, moved into the suburbs. Developers advertised lower taxes, spacious living with no industrial pollution, and better family life. They were right at first, but once large numbers of city inhabitants and businesses joined the suburbs, they became noisy, crowded, and were taxed even higher to provide services already in place in the city.

Because of the land sprawl, people could no longer walk to the store or work. Congestion and traffic thickened, resulting in jams lasting hours. People with city jobs had to sit in traffic for 10 to 15 hours or more weekly. When fuel prices soared, cities scrambled to set up mass transit. Some met their populations' needs, but as populations grew, subways and city trains far exceeded their original carrying capacity.

Increased driving has also impacted suburbanite health. Those sitting for long hours in cars and at jobs get little exercise. When people lived in cities and walked everywhere, obesity and heart disease rates were much lower.

Transportation

Urban and suburban living may seem like a good option, but moving millions of people in and out of a city daily without mass transit is a huge task with a negative environmental impact. Millions of vehicles carrying one person cause traffic jams and increase air pollution.

Bus systems, subways, and monorail transportation have eased congestion and pollution, but not all cities can afford modern transportation. Cities successful at reducing urban sprawl started before their populations got huge. Unfortunately, cities like Houston, Texas (population nearly $4^1/2$ million), face tremendous costs and construction nightmares in converting from an historically one-person, one-car mentality to subway and light-rail transportation.

In cities such as Portland, Oregon, however, where bike routes are common, light-rails have been constructed, zoning has been enacted, and developers are required to invest in established neighborhoods, environmental sustainability works.

Other forward-thinking cities augmented existing federal highways with green alternatives. Many Canadian coastal cities use large vehicle-carrying ferries to transport people and cars to local destinations. Cities such as Venice, Italy, use canals as tourist attractions as well as for transportation.

Water

Land use is also important in other ways. Its impact on the hydrologic cycle and on soil and water quality is direct, as well as indirect. Direct impacts increase soil erosion, flood, drought, and river and groundwater changes.

Indirect effects come from the environmental impacts of land use, and the subsequent climate and environmental changes. For example, land cover can limit water available for groundwater recharge, reducing groundwater discharge to rivers. These hydrologic cycle effects can greatly impact downstream flow.

Ecological Footprint

In the past few centuries, more and more people have moved to towns, cities, or other centrally developed areas. For example, only 15% of U.S. citizens lived in cities during the Civil War. Today it is over 80%. Urban sprawl creates problems in feeding, transporting, sheltering, providing water, and disposing of waste from all those people.

A population's *ecological footprint* is the amount of surface area needed to provide for its needs and dispose of its waste.

Americans have a large ecological footprint, around 10 hectares per person. (*Note*: One hectare = 10,000 square meters [~ 2.5 acres].) Other developed countries such as Argentina (2.5 hectares/person), Sweden (5.0 hectares/person), and Australia (7.5 hectares/person) have smaller footprints. The world's average ecological footprint is estimated at 2.2 hectares/person.

A mathematical equation, called IPAT, is used to calculate human impact on the environment:

$$I = P \times A \times T$$

The equation solves for I (total impact) by using P (population size), A (affluence), and T (technology level), and provides an idea of a specific population's influence on its local and regional resources.

Loss of Coastal Habitat

Since over half of the world's population currently lives within 60 km of coastal waters and this number is expected to double within the next three to four decades, the oceans will definitely feel an impact. Six of the world's eight largest cities (population over 10 million) are coastal. Moves to coastal areas are driven by poverty and affluence. Low-income people move to cities for jobs, while wealthy people expand shoreline development for resort hotels and seaside homes. The World Resources Institute estimates about half of the world's coastal ecosystems are threatened by development, with most located in northern temperate and equatorial regions, including the coastal zones of Europe, Asia, the United States, and Central America.

Water treatment technology has improved, leading to further growth in developed areas where the new technology is used. However, many developing countries cannot afford to implement these new technologies, and population growth in coastal cities continues to overwhelm existing waste treatment systems. Disease from contaminated water creates a limiting factor on many species' growth in coastal regions.

Population growth also affects the world's oceans. Many interacting factors add to the degradation of marine ecosystems and loss of biodiversity. Population impacts on ocean resources must be better understood if resources are to be sustained without exceeding carrying capacity.

Public and Federal Lands

The world's land area covers around 133 million km^2 (56 million mi^2) or nearly 30% of the planet. Grassland, agriculture, and forests cover around 65% of this. Regions that don't fall into the three main categories are made up of desert, tundra, wetlands, brush, urban areas, ice, snow, and bare rock. Figure 14.1 shows global land use as estimated by the Food and Agriculture Organization.

Not all of this land can be easily or freely utilized. Nearly 10% is currently protected in wildlife refuges, national forests, parks, and nature preserves. Over 11% is

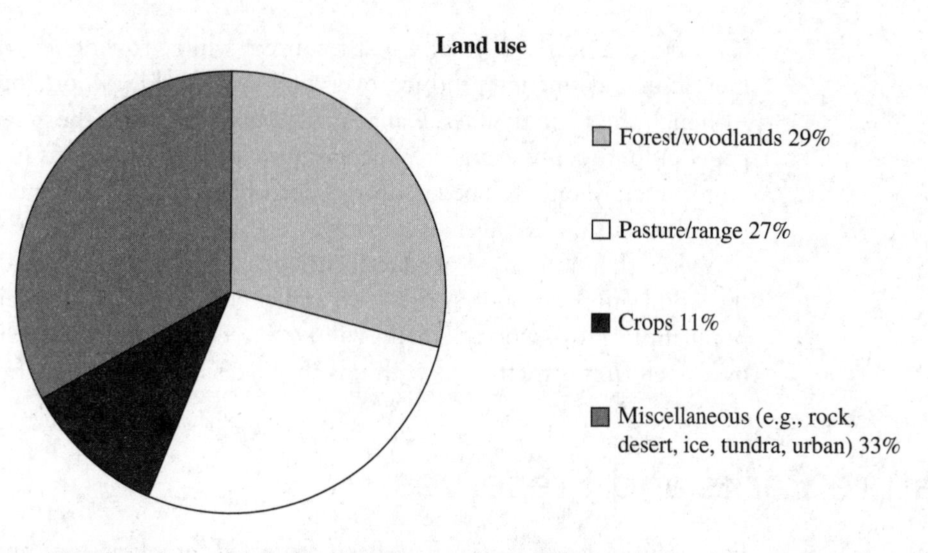

Figure 14.1 Globally, land is used in different ways depending on culture, economics, soil, and climate.

under cultivation for crops. However, many countries are expanding agriculture with new methods for feeding ever-growing populations.

This is nothing new. Billions of acres of forests have been put to the ax over the past 10,000 years in central Asia and Europe, while continents such as South America are only recently feeling the pressure to use more of their forests.

Communal Property

A resource management problem is that we all share global resources (e.g., air, wildlife, mountains, ocean fish). Unfortunately, these can be exploited by a single person, group of people, or nation.

In 1968, biologist Garret Hardin wrote, "The Tragedy of the Commons," an article explaining how commonly held resources are often misused through self-interest. He thought since everyone shared the same resources and need for those resources (e.g., groundwater), it made it easier to disclaim responsibility for misuse of the resource.

Current ecologists say an *open-access system,* where there are no rules regulating a resource, makes management difficult. Instead they suggest a *communal resource management system* where (1) there are clear boundaries, (2) people live on the land a long time, (3) community size is regulated, (4) people must work together to preserve scarce resources, (5) people have input into rule setting, (6) resource use is actively monitored, (7) conflict management is in place, and (8) compliance incentives exist. When an entire community is responsible for an area or resource, they work together to sustain it, to the benefit of all.

Forests

Although over half of the planet's original forests are gone, the 3.8 billion hectares left cover nearly 30% of the global land surface. In fact, nearly every industry uses wood or wood products in manufacturing or marketing activities. Worldwide wood consumption has almost doubled in the past 50 years.

Timber is used for building lumber, plywood, particle board, chipboard, and veneer, about one-half of the world's wood consumption. As with other resources, developed countries use nearly 80% of wood resources while growing less than half of that amount. Developing countries produce over half the world's wood, but only use about 20%. Although the United States, Canada, and Russia produce the greatest amount of wood and paper pulp, they are using ever-increasing sustainable methods such as replanting cut areas. Unfortunately, undeveloped countries don't have the governmental requirements or funding to replace cut timber in sustainable ways.

More than just wood products, such as paper and building materials, fully one-third of the world's population uses firewood and charcoal for heating their homes and cooking. As populations grow, wood demand will far exceed dwindling supplies. Additionally, cooking over open fires is inefficient with less than 10% of the heat energy actually being used.

Nature Parks and Preserves

For centuries, natural areas have been protected for religious or hunting reasons or as a playground for royalty. Only recently, in the past 100 years or so, have some areas been set aside to protect the environment and native species.

Yosemite, in California, was the first area in the United States set aside during the Civil War to protect nature. President Abraham Lincoln deeded it to the state of California, since there was no national office to care for that area then. In 1890, Yosemite was given back to the federal government as a national park. President Ulysses S. Grant designated 800,000 hectares (about 2 million acres) of Wyoming, Montana, and Idaho territories as the first national park (Yellowstone) in the world in 1872. Yellowstone was established to protect the natural curiosities and wonders encompassed in geysers, hot springs, and canyons. Yellowstone has 300 million visitors yearly.

In 1901, President Theodore Roosevelt established 51 national *wildlife refuges* in the United States. The number of these areas now totals 540, comprising nearly 40 million hectares of land for the protection of species. The latest addition was made by President Jimmy Carter when he created the Alaska National Interest Lands Conservation Act in 1980 (protecting about 22 million acres).

Since the establishment of the first parks, the U.S. national park system has expanded to 388 parks, monuments, historic sites, and other areas totaling 280,000 km^2 (108,000 mi^2). Canada has over 1,470 parks and protected areas enclosing around 150,000 km^2.

Restoration Ecology

Unfortunately, not everyone agrees costly parks and preserves are the best way to protect nature. Some herds like the elk in the Grand Teton National Park have increased so much that park officials have reintroduced wolves, a natural predator, into the park to keep the herds from overpopulating and starving. Local residents are not excited about having wolves nearby.

Controversy over wilderness areas, wildlife refuges, nature preserves, and parks revolves around whether countries are protecting natural resources, species, and ecosystems, or whether they are setting aside sightseeing and recreational areas.

The *1964 Wilderness Act* defined wilderness as an area of undeveloped land affected primarily by the forces of nature, where man is a visitor who does not remain; it contains, ecological, geological, or other features of scientific or historic value; it possesses outstanding opportunities for solitude or a primitive and unconfined type of recreation; it is an area large enough so that continued use will not change its unspoiled natural conditions.

Environmentalists want more undisturbed areas designated as wilderness areas. Miners, loggers, and ranchers want fewer areas set aside. The rationale for wilderness areas is that they provide (1) wildlife refuges and reproductive areas, (2) a research basis for species' changes, (3) a place of solitude and primitive recreation (no motor vehicles), and (4) an area of undisturbed natural beauty for future generations.

Water pollution has become a problem in over 75% of all U.S. refuges. Energy activities (e.g., oil and gas drilling) on the north slope of Alaska's Brooks Range is an example of a hotly disputed extension into a natural area. Environmentalists foresee ecosystem destruction and toxic pollution like that which resulted from the Exxon Valdez oil spill, compromising protected waterways and species.

Marine Preserves

Globally, fish stocks are decreasing from overfishing. Although many nations have recognized this serious issue and have policies protecting their shores, others have not.

Biologists report that "no take" refuges average twice as many species as surrounding areas, and individuals are 30% larger. To protect depleted marine resources, ecologists urged nations to protect at least 20% of coastal shore territories as marine refuges.

Coral reefs are especially at risk. Remote sensing currently shows living coral reef over approximately 285,000 km^2 (110,000 mi^2), which is less than half prior estimates. In fact, scientists report 90% of all reefs are threatened by sea temperature change, coral mining, sediment runoff, ocean dumping, and destructive fishing methods. With current trends, researchers predict living coral reefs will be gone by 2060.

Conservation Methods

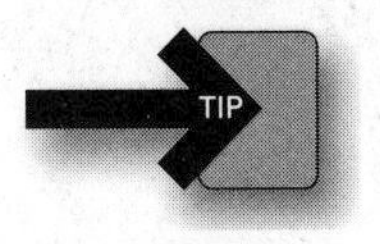

Worldwide, there are over 100,000 protected areas enclosing nearly 19 million km^2 of biological habitat. The *International Union for the Conservation of Nature* (IUCN) created a world conservation strategy to (1) maintain critical ecological processes (soil and nutrient recycling and water purification), (2) preserve genetic diversity in plants and animals, and (3) ensure sustainability of wild species and ecosystems.

Forest protection, management, and establishment of national parks and preserves are often accomplished via economics. Environmental organizations (e.g., Nature Conservancy, World Wildlife Fund, and Conservation International) buy bank debts and work with developing countries to forgive debts if they strive to protect important biological regions. These debt-for-nature swaps have been done in a number of countries, including Bolivia, Costa Rica, the Dominican Republic, Guatemala, Panama, and Peru.

The IUCN developed a world conservation strategy for natural resources. It focuses on the following three objectives: (1) preserving ecological processes essential to human life and development (e.g., nutrient recycling, soil regeneration, and water purification), (2) protecting genetic diversity of plants and animals, and (3) upholding sustainability of wild species and ecosystems.

In 1986, the United Nations Educational, Scientific, and Cultural Organization (UNESCO) came up with a plan to merge human and wildlife needs in protected areas. The *Man and Biosphere (MAB)* program divides protected areas into different use zones. A core zone is reserved for crucial ecosystems and endangered wildlife. Very limited scientific study is allowed in the core. An outer pristine buffer zone permits some research facilities and ecotourism. Beyond this, zones for sustainable harvesting, a multiuse area, and permanent housing are allowed.

An example of a MAB reserve is the Sian Ka'an Reserve south of Cancun, Mexico. The core includes 528,000 hectares (1.3 million acres) of coral reef, bays, marshes, and tropical forest. Over 335 bird species, manatees, jungle cats, monkeys, and sea turtles reside there. The local economy gets a boost from tourism, farming, and lobster fishing. Community leaders protect the reserve, while also working to improve local living standards.

Mining

Mining involves the extraction of economically important minerals and metals by different methods. Most economic minerals have high metal levels (e.g., aluminum, copper). Nonmetallic resources (e.g., diamond, graphite) are often sought for practical or aesthetic reasons.

Minerals

Most rocks are composed of *minerals* with an assorted combination of differing chemical elements.

A *mineral* is a naturally found, inorganic solid with a specific crystalline structure and chemical composition.

The chemical structure of minerals is exact, or slightly varied within limits. Minerals have specific crystalline structures and belong to different groups according to the way the mineral's atoms are arranged. Elements like gold, silver, and copper are found naturally and considered to be minerals.

Crystalline Structure

TIP

Most minerals have a crystalline form with specific geometric arrangements. These structures can be the same between different minerals, but their chemical makeup is different. A mineral's internal structure determines its physical and optical properties, shape, hardness, cleavage, fracture lines, specific gravity, refractive index, and optical axes. The regularly occurring arrangement of atoms and molecules in space determines form.

Geologists have identified over 3,000 minerals all over the world. Australia has particularly rich sources of aluminum, lead, and zinc, while Russia and Canada are rich in nickel. The United States is a prime source of copper.

Mineral Formation and Extraction

Minerals are found in deposits formed from evaporation, wind deposit, sedimentation, or ancient volcanic heating. Depending on a deposit's depth, minerals are extracted through open-pit mines, *strip mining*, or deep underground mines. The California Gold Rush of the mid-1800s saw thousands of miners trying *placer mining* (i.e., using water to wash away soil from nuggets in shallow pans or in long wooden troughs). Today, placer mining in Canada and Alaska is done by blasting hills with water cannons to separate gold and other minerals. This method uses tons of water and clogs streams with sediment.

Mining Impacts

Other mining methods remove and crush huge amounts of soil and rock to obtain desired metals. Cyanide, mercury, and other hazardous elements are often used during extraction, along with large amounts of water to wash away extra rock. Consequently, ore removal and purification causes water pollution through postprocessing contamination by arsenic, heavy metals, and acids. Mining runoff pollutes streams, lakes, and reservoirs, killing marine life and poisoning ecosystems.

The EPA lists over 100 toxic air pollutants released yearly from U.S. mining operations. Much of this pollution comes from *smelting* or cooking ore to release its metals. In the past 150 years, wood fires have been used to extract copper from ore. This process released sulfur dioxide gas, killed plants, and acidified soil for miles.

Heap-leach extraction, where ore piles are sprayed with a weak cyanide solution to extract gold, is another bad idea. Costing the EPA millions of dollars for cleanup, one abandoned mine in Colorado had tons of remaining mine waste in addition to large, leaking cyanide ponds.

Mining Legislation

The 1977 *Surface Mining Control and Reclamation Act* mandated restoration of strip-mined land. Though expensive (i.e., $10,000/hectare) and difficult due to the amount of acidified soils, progress has been made. Additionally, the *Clean Air Act of 1970* began regulating many toxic and environmentally harmful mining waste compounds.

In addition to strict regulations, community recycling mandates lower the need for new mining. For example, recycling waste aluminum requires one-twentieth of the energy of extracting new aluminum. Today over 65% of aluminum drink cans are recycled.

Review Questions

Multiple-Choice Questions

1. Urbanization came about from

(A) less mass transit
(B) change in growing seasons
(C) lack of high-rise properties
(D) gradual changes in land use and societal culture
(E) lower city taxes

2. Push and pull immigration factors include

(A) overpopulation
(B) the arts
(C) lack of country resources
(D) communications
(E) all the above

3. The amount of surface area needed to provide for a population's needs and dispose of its waste is called its

(A) niche
(B) tillage area
(C) ecological footprint
(D) growth index
(E) urbanization

4. When ore piles are sprayed with a weak cyanide solution to extract gold, it is known as

(A) strip mining
(B) deep shaft drilling
(C) open-pit mining
(D) placer mining
(E) heap-leach extraction

5. All of the following affect a mineral's internal structure except

(A) cleavage
(B) boiling point
(C) shape
(D) hardness
(E) optical axes

6. Deposits formed from evaporation, wind deposit, sedimentation, or volcanic heating contain

(A) minerals
(B) water
(C) lava
(D) algae
(E) radioactive tracers

7. The Surface Mining Control and Reclamation Act mandated

(A) a 20% decrease in sulfur emissions
(B) no new mining without smokestack filters
(C) clear air
(D) restoration of strip-mined land
(E) recycling of aluminum

8. Megacities have over

(A) 1 million inhabitants
(B) 4 million inhabitants
(C) 5 million inhabitants
(D) 10 million inhabitants
(E) 20 million inhabitants

9. A naturally found, inorganic solid with a specific crystalline structure and chemical composition is called a

(A) rock
(B) fossil
(C) mineral
(D) biome
(E) carbon nanotube

10. What percent of urban sewage is treated in Latin America?

(A) 1%
(B) 2%
(C) 3%
(D) 5%
(E) 10%

11. Bike routes, light-rail, zoning, and development regulations increase

(A) cultural diversity
(B) exercise options
(C) taxes
(D) corporate options
(E) environmental sustainability

12. Direct impacts on the hydrologic cycle include all the following except

(A) flood
(B) drought
(C) mining
(D) river flow
(E) groundwater reservoirs

13. Feeding, transporting, sheltering, providing water, and waste disposal are all problems of

(A) small countries
(B) rural living
(C) urban sprawl
(D) island nations
(E) Arctic scientists

14. Grassland, agriculture, and forests

(A) provide visual appeal
(B) use very little water
(C) are all planted by humans
(D) cover the world's land area
(E) are always found in great amounts worldwide

15. When ore is heated to extract minerals, it is known as

(A) deep shaft mining
(B) smelting
(C) placer mining
(D) a sustainable method of extraction
(E) ablation

16. What percent of Mexico City's population lives in shantytowns built on undeveloped land from scavenged materials?

(A) 10%
(B) 25%
(C) 35%
(D) 50%
(E) 70%

17. An unforeseen result of suburban living is

(A) less traffic
(B) less exercise
(C) closer community ties
(D) lower taxes
(E) greater vehicle wear and tear

18. When water is used to wash away soil from gold nuggets in shallow pans or in long wooden troughs, it is called

(A) strip mining
(B) deep shaft drilling
(C) open-pit mining
(D) placer mining
(E) heap-leach extraction

19. A communal resource management system includes all the following factors except

(A) clear boundaries
(B) active monitoring of resource use
(C) people living on the land a long time
(D) privatization and self-interest
(E) people having input into rule setting

› Answers and Explanations

1. D—As populations and cities grew there were more jobs in the cities.

2. E

3. C—The smaller the ecological footprint, the better the use of resources.

4. E—This leaching releases sulfur dioxide and acidifies soil and plants for miles.

5. B—Internal structure may have an effect on boiling point, but not the other way.

6. A—Rocks and dust particles are made of various minerals.

7. D

8. D

9. C—Minerals are primarily characterized by crystalline structure and hardness.

10. B

11. E—By using many options, the drain on one resource is less.

12. C—Mining may have a small effect in an area but not on the overall hydrologic cycle.

13. C—Too many people in a small area causes shortages of resources.

14. D—Vegetation on the Earth's surface takes these different forms.

15. B—Smelting causes pollution with its extraction chemicals as well as its waste.

16. D

17. B—Less exercise has also resulted in unhealthy weight gain.

18. D—A placer is a glacial or water deposit of gravel or sand containing heavy minerals.

19. D—Privatization encourages selfishness and fewer community bonds between people working for the greater good.

Free-Response Questions

1. The IUCN divided protected areas into five categories with various levels of protection and human impact. Answer the following questions using the given table.

IUCN categories of protected areas.

	CATEGORY	PERMITTED HUMAN IMPACT OR INTERVENTION
1	Ecological reserves/wilderness areas	Minimal to none
2	National parks	Minimal
3	Natural monuments/archaeological sites	Minimal to medium
4	Habitat/wildlife management areas	Medium
5	Cultural/scenic landscapes, recreational areas	Medium to high

(a) What differences in transportation would you expect to see between "Cultural/scenic landscapes, recreational areas" and "Ecological reserves/wilderness areas"?

(b) What is the correlation between the category of protected area and permitted human impact or intervention and why?

2. Humans have looked for minerals and ore for centuries. Besides the hazards to the environment, mining is dangerous to miners as well. Minerals are found in deposits formed from evaporation, wind deposits, sedimentation, or ancient volcanic heating. Depending on a deposit's depth, minerals are extracted in various ways.
   (a) Describe different mining methods.
   (b) Explain three major impacts on the land from mining.

Free-Response Answers and Explanations

1.

a. In cultural and scenic areas, roads are generally provided for visitors to enjoy the surroundings. Roads necessarily alter the landscape, while human activity due to roads and motorized vehicles further alters the landscape, air, and water quality. Because reserves and wilderness usually strictly prohibit motorized vehicles, road construction is minimal while human transportation is done in an environmentally friendly way. The transportation differences speak directly to the use humans intend to make, or not to make, of the areas in question.

b. The more protected an area is, the less human intervention is permitted. Though humans are also fellow organisms, our ability to reshape the environment according to our own needs far exceeds that of all other species, marginalizing and endangering many of them in the process. Ironically, to conserve the natural beauty and bounty that attracts so many urban dwellers to "wild" areas, ecological reserves and other like areas must be protected from the very organisms that wish to enjoy it.

2.

a. Open-pit mines, strip mining, or deep underground mining methods are used depending on the type of mineral or ore sought. Simple placer mining of the mid-1800s (i.e., using water to wash away soil from nuggets in shallow pans) has been replaced today by blasting hills with water cannons to separate gold and other minerals.

b. Ore removal and purification causes water pollution through postprocessing contamination by arsenic, heavy metals, and acids. Mining runoff pollutes streams, lakes, and reservoirs killing marine life and poisoning ecosystems. Smelting or cooking ore over wood fires to release its metals (e.g., copper extraction) releases sulfur dioxide gas, kills plants, and acidifies soil for miles.

❯ Rapid Review

- A population's ecological footprint is the amount of surface area needed to provide for its needs and dispose of its waste.
- Urban populations increase naturally by more births than deaths and by immigration.
- Immigration is affected by different characteristics or push and pull factors.
- In Mexico City, roughly 50% of the population lives in shantytowns built on undeveloped land from scavenged materials.
- In densely populated cities such as Jakarta, Indonesia, a worker spends around three to four hours traveling to and from work each way due to traffic jams.
- Millions of vehicles carrying one person cause traffic jams and increase air pollution.

- Bus systems, subways, and monorail transportation have eased congestion and pollution, but not all cities can afford modern transportation.
- The world's land area covers around 133 million km^2 (56 million mi^2), or nearly 30% of the planet. Grassland, agriculture, and forests cover around 65% of this area.
- Over half the world's population lives within 60 km of coastal waters, and this population is expected to double within the next three to four decades.
- Yosemite was the first area in the United States set aside by President Abraham Lincoln during the Civil War to protect nature.
- Developed countries use nearly 80% of global wood resources while developing countries produce over half the world's wood, but only use about 20%.
- President Theodore Roosevelt established 51 national wildlife refuges in the United States. There are now 540, which surround nearly 40 million hectares of land.
- Ninety percent of all reefs are threatened by sea temperature change, coral mining, sediment runoff, ocean dumping, and destructive fishing methods. Researchers predict living coral reefs will be gone by 2060.
- The International Union for the Conservation of Nature created a world conservation strategy to (1) maintain critical ecological processes, (2) preserve genetic diversity, and (3) ensure sustainability of wild species and ecosystems.
- The Man and Biosphere program calls for the division of protected areas into different use zones, with a core area reserved for crucial ecosystems and endangered wildlife.
- Depending on the deposit depth, minerals are extracted through open-pit mines, strip mining, or deep underground mines.
- A mineral is a naturally found, inorganic solid with a specific crystalline structure and chemical composition.
- A mineral's internal structure determines its physical and optical properties, shape, hardness, cleavage, fracture lines, specific gravity, refractive index, and optical axes.
- The 1977 Surface Mining Control and Reclamation Act mandated restoration of strip-mined land.

CHAPTER 15

Energy Consumption, Conservation, and Fossil Fuels

IN THIS CHAPTER

Summary: This chapter will examine some of our historical and current fuel choices, while exploring in greater detail the impacts of consumption, conversion, and use.

Keywords

✪ Work, energy, calorie, power, joule, conversion, laws of thermodynamics, Hubbert's peak, oil shale, tar sands

Energy

All organisms need energy to function. More is needed for species-related or cultural activities (e.g., migration, hibernation, fight-or-flight response, finding food and shelter, or reproduction). Just as there are different energy uses, energy comes in different types. For example, people get energy from the compounds (e.g., proteins, carbohydrates, and fats) stored in food. A *calorie* is a unit of food energy.

> One *calorie* unit equals the amount of energy needed to heat 1 gram of water to 1 degree Celsius.

Energy provides the power to do work. *Work* is defined as force exerted over distance. *Power* is the rate of flow of energy or the rate at which work is accomplished. Another way to measure energy is to measure force and work. A *newton* describes the force needed to accelerate 1 kilogram by 1 meter per second. A *joule* is the amount of work accomplished when a force of 1 newton is performed over 1 meter or 1 ampere per second travels through 1 ohm (a unit of electricity). Several units of energy are listed on page 273.

Laws of Thermodynamics

To understand energy, you also need to know the laws of thermodynamics. These affect all types of energy, their conversion, and storage. Here they are:

1. The *first law of thermodynamics* states energy can be neither created nor destroyed. It is simply transported or changed into another form. For example, deep ocean vents have communities of microorganisms that use sulfur compounds released from thermal vents for chemosynthesis in the same way plants use sunlight for photosynthesis.

2. The *second law of thermodynamics* describes how the universe tends toward *entropy* (chaos, disorder, or randomness). The original energy amount is no longer available in its original form but has changed to another form. For example, much energy produced by a power plant is lost through heat in turbines and when sent through electrical lines.

Conversions

TIP

One of the big energy concerns is the amount of energy needed during energy refining and production. In fact, processing accounts for nearly half of all energy lost during conversion to more usable forms, transportation, or use.

For example, when coal is used to produce electricity, nearly 65% of the original energy is lost during thermal conversion at the power plant. Another 10% is lost in electrical transmission and voltage changes for household use. Most losses are seen during fossil fuel refining. About 75% of oil's original energy is lost during distillation into gasoline and other fuels, transportation to market, storage, and engine combustion.

Natural gas has much less waste since it needs little refining. It is transported through underground pipelines and burned with 75–95% efficiency in regular and high-efficiency furnaces. It also contains more hydrogen-to-carbon atoms, and so produces much less carbon dioxide (about one-half less compared to oil or coal), reducing its impact on global warming.

Fossil Fuels

Fossils fuels are hydrocarbons formed into coal (solid), oil (fluid), and natural gas (mostly methane). These can be used as fuels by themselves or processed to produce purer products like propane and gasoline. Fossil fuels (oil and natural gas) are also used in the petrochemical industry to make chemicals, plastics, and fertilizers.

> *Fossil fuels* are solids, liquids, and gases created through the compression of ancient organic plant and animal materials in the Earth's crust.

Fossil fuel burning is the biggest single source of human-created air pollution in the industrialized world. Reducing the amount of smoke, ash, and combustion products from fossil fuel burning is critical to the future of life on Earth.

Energy Production and Use

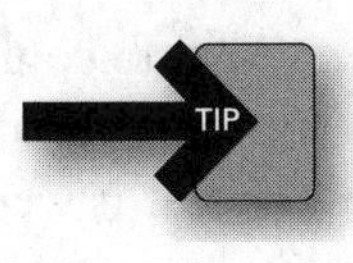

Fossil fuels produce around 82% of the energy used in the United States, with 45% in the form of gasoline and 29% as heating oil and diesel fuel. In 2012, coal was used as fuel for 37% of the 4 trillion kilowatthours of electricity generated in the United States. Nuclear power provides roughly 20% of U.S. electricity, natural gas provides 27%, and other alternative fuels provide about 14%.

Fossil fuels power vehicles, industrial manufacturing, and homes and businesses. Of U.S. energy consumption, 45% is used in the mining, smelting, and forging of metals, as well as the manufacture of plastics, solvents, lubricants, fertilizers, and organic chemicals. Homes and businesses use about 11% in lighting, heating, air conditioning, cooking, and water heating. Transportation uses another 28%.

Hubbert's Peak

Fossil fuels have been exploited worldwide for decades following many technological advances. However, it wasn't until large cities became sooty, dirty places from fossil fuel burning that people began to think there might be a better way. Scientists debated whether it was easier to stick with fossil fuels and its problems or switch to a better energy source. In 1956, Shell Oil Company geophysicist M. King Hubbert calculated that the oil well extraction rate in the United States (lower 48 states) would peak around 1970 and begin dropping from then on. At the time, people didn't believe him. Hubbert was criticized by oil experts and economists, but it turned out he was right. Oil production peaked at around 9 billion barrels/day in 1970.

China, which recently became the world's largest energy consumer, is projected to consume more than twice as much energy as the United States by 2040.

Since the industrialized world depends on fossil fuels, experts are trying to figure out how many years are left before all known fossil fuel (coal, oil, and natural gas) reserves are gone, based on world consumption rates. Currently, around 65% of electricity produced comes from fossil fuels worldwide, but that number is expected to shrink as supplies dwindle and alternatives expand.

Fossil Fuel Resources

Even if we use *all* available fossil fuels globally, there will always be some left in the Earth. Liquid oil comes out readily, but oil trapped between rock layers or in honeycomb configurations is not easily extracted. In fact, a 30–40% yield from an oil formation is common. There are technological as well as cost limitations to oil drilling. Getting more oil from a drilled deposit using different methods is known as *secondary recovery.*

Currently, oil production in the United States from all sources is 8.5 million barrels a day. Increased by hydraulic shale fracturing, U.S. oil production is expected to average over 9.3 million barrels in 2015, according to the Department of Energy. This can be compared to 10 million barrels per day in Saudi Arabia from only 750 wells. Nearly two-thirds of the world's *proven* oil reserves are located in the Persian Gulf countries of the Middle East.

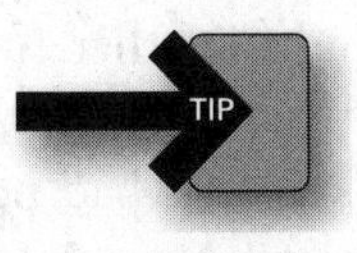

Geologists, performing calculations with updated global oil data, found global oil production will peak between 2014 and 2018 depending on demand and introduction of alternative fuels. Of the approximately 2.5 trillion barrels of the Earth's total recoverable supply of oil (*United States Geological Survey* estimate), over 50% of it has already been used. Since fossil fuel consumption is still rising and proven reserves are less than 1 trillion barrels, worldwide supply is estimated to last between 30 to 40 years at current rates.

Surprisingly, Canada is the biggest supplier of oil to the United States, with Saudi Arabia a close second.

Fossil Fuel Disadvantages

It is estimated that around 270 billion cubic meters of *tar sands* in northern Alberta could also provide over 2 billion barrels of oil per day to the United States in the next 5 years. This alternative oil resource could double the total North American oil reserves if it were possible to recover them without steep environmental impacts. However, unless better technology is found to reduce toxic sludge and water pollution caused during tar sand processing, public outcry will keep this resource from development. Similar concerns have been raised about U.S. *oil shale,* a sedimentary rock that contains solid organic matter called *kerogen.* When heated to 480°C (900°F), kerogen melts and can be drawn up out of the ground. Oil shale is located in Arkansas, Colorado, and Wyoming, as well as some eastern states. This makes the huge amounts of water needed for extraction a costly process, in addition to generating huge amounts of waste and pollution.

Another hotly debated area with large oil reserves is Alaska's Arctic National Wildlife Refuge (ANWR). This area, which contains millions of geese, swans, shorebirds, migratory birds, polar bears, arctic foxes, arctic wolves, and the largest herd of caribou in the world (130,000), is estimated to contain as much as 12 billion barrels of oil. Petroleum companies claim they can extract the oil with little environmental impact, but ecologists cite buried waste, heavy machinery impacts, and pipeline and drilling spills on the tundra as major negatives.

Native peoples are divided in their support of ANWR drilling. Those who work in the oil industry generally favor it, while others who support traditional ways and need caribou as a primary food source oppose it.

Conservationists agree with federal ANWR drilling advocates that we need to decrease U.S. dependence on foreign oil. However, they prefer vehicles with greater fuel efficiencies, and the use of alternative energy sources (e.g., wind and solar) to fill the gaps rather than the sacrifice of pristine environments.

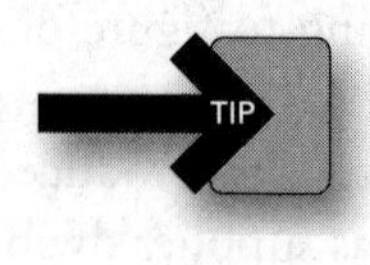

The time for energy alternatives is now. Just like driving a car, you can't wait until you run out of gas to stop at a station and refuel. Fossil fuel depletion could be slowed if demand could be lowered by switching to new, more efficient energy resources.

Oil

Humans have used oil since the ancient Chinese and Egyptians burned oil for lighting. In 1839, Abraham Gesner, a governmental geologist in New Brunswick, Nova Scotia, discovered *albertite* (a solid coal-like material). After immigrating to the United States, Gesner developed and patented a process for manufacturing *kerosene.* He is often called the father of the petroleum industry.

Before the 1800s, however, light was produced from torches, tallow candles, and lamps burning animal fat. Because it burned cleaner than other fuels, whale oil was popular for lamp oils and candles. However, it was expensive. A gallon cost about $2.00, which today would be around $760 per liter ($200 per gallon). And we think our oil is expensive!

As whale oil became more and more expensive, people started looking for other fuel sources. In 1857, Michael Dietz invented a clean-burning kerosene lamp. Almost overnight, whale oil demand dropped. Most historians and ecologists believe that if kerosene had not come onto the market, many whale species would be extinct from overhunting.

Kerosene, known as "coal oil," was cheaper, smelled better than animal fat when burned, and it didn't spoil like whale oil. By 1860, around 30 U.S. kerosene plants were in production. Kerosene was used to light homes and businesses before the invention of the electric lightbulb by Thomas Edison in 1879.

Oil Wells

Although the first oil well in North America was drilled in Oil Springs, Ontario, in 1855, U.S. oil discovery peaked in 1930 with the discovery of the Spindletop, East Texas, field, which produced 80,000 barrels of oil per day.

For decades, industrialized, developed countries with only 20% of the world's population used around 80% of all the commercial energy; the rest of the world's population used the remaining 20%. However, with progress in countries like China and India, energy use is increasing substantially. It is estimated that by 2040, China and India will more than double their current energy use to 34% of the projected total world energy consumption.

Oil Demand

Worldwide consumption of liquid fossil fuels rose to 90 million barrels of oil per day in 2014, it is expected to reach nearly 96 million barrels of oil per day by 2016, and over 112 million barrels of oil per day in 2030.

Americans use nearly 20% more transportation oil than we did 40 years ago. Even though today's vehicles are more fuel efficient, there are 50% more vehicles on the road today than in the 1970s.

Coal

Like liquid and gaseous fossil fuels, coal is derived from decayed organic plant material formed and compacted by geological forces into a high-carbon-containing fuel. At reduced rates of use and population growth, this may last 500 years. However, extractable *proven* reserves will last btween 110–130 years at current growing consumption rates.

The reason so many people consider coal to be a great fuel source is because global coal deposits are about 10 times greater than conventional oil and gas resources added together. Coal seams in rock can be 100 feet thick (around 35 meters) and stretch thousands of square miles across huge areas that were originally tropical rain forests in prehistoric times.

Coal deposits that have been mapped, measured, and found to be economically recoverable are known as *proven reserves. Known reserves* are coal deposits that have been identified but not well mapped, while *ultimate reserves* include known and unknown coal deposits.

Coal Demand

In 2014, coal provided 42% of the world's electricity. The IEO estimates coal will increase by 2% per year from 2009 to 2015 and will account for over one-third of total world energy consumption in 2030.

Global coal reserves are *estimated* at over 860 billion tons. At current rates of use and population growth, this may last 1,000 years. However, extractable *proven* reserves will last roughly 118 years at current growing consumption rates.

About a quarter of known coal reserves are in the United States. Many electricity-generating plants burn coal, and many people consider coal the nation's backup energy source.

Unfortunately, coal burning creates a lot of atmospheric and particulate pollution. In fact, coal combustion accounts for nearly 25% of all atmospheric mercury pollution in the United States today. With current technology, coal energy appears to be a win–lose situation.

Extraction and Purification Disadvantages

The major drawback to coal is that its extraction from the Earth is extremely dangerous. Thousands of miners are killed every year in mining cave-ins and other accidents. Coal dust, thick in the mining process, is also very toxic to the lungs, causing *black lung disease* and other debilitating respiratory diseases. Though lessened with advanced respirators, black lung is a major occupational hazard for coal workers.

Strip or surface coal mining has fewer direct hazards to workers, but a huge impact on the local environment. The removal and processing of rock makes the area unfit for any other use. Prior to mine reclamation laws, mine drainage and acidic runoff from coal tailings were toxic by-products of strip mining. Thousands of miles of streams and creeks were polluted from coal mining operations. An additional insult to the Appalachian environment comes from the dumping of excess strip mining soil and debris into valleys, rural streams, farms, and other sites.

Coal is often contaminated with sulfur, sometimes as much as 10% by weight. This must be removed by washing or flue-gas scrubbing, or it will be released during burning to form sulfur dioxide (SO_2) or sulfate (SO_4). Burning releases or reacts to form around 18 million metric tons of SO_2, 5 million metric tons of nitrogen oxides (NO_x), 4 million metric tons of airborne particulates, 600,000 metric tons of hydrocarbons and carbon monoxide (CO), and nearly a trillion metric tons of carbon dioxide (CO_2). These atmospheric releases account for 65% of the SO_2, 50% of the industrial CO_2, and 30% of the NO_2 released yearly in the United States. These and other pollutants negatively impact the environment, human and animal health, plants, and buildings.

The good news is that sulfur can be removed before coal is burned and nitrous oxide reactions can be limited. However, government policy, as well as economic and technological incentives will have to be strengthened and supported to a much greater degree to motivate change.

› Review Questions

Multiple-Choice Questions

1. What percentage of United States energy comes from fossil fuels?

   (A) 10%
   (B) 18%
   (C) 37%
   (D) 62%
   (E) 85%

2. Coal is often contaminated with what element?

   (A) Carbon
   (B) Silicon
   (C) Sulfur
   (D) Uranium
   (E) Sodium

3. The amount of energy needed to heat 1 gram of water to 1 degree Celsius is a

   (A) gigabyte
   (B) pound
   (C) liter
   (D) calorie
   (E) nanosecond

4. Solids, liquids, and gases created through the compression of ancient organic plant and animal material in the Earth's crust are called

   (A) biomass
   (B) inorganics
   (C) sustainable fuels
   (D) nuclear fuels
   (E) fossil fuels

5. About 75% of the original energy in oil is lost during distillation into

   (A) kerosene
   (B) gasoline
   (C) jet fuel
   (D) road tar
   (E) ethanol

6. Coal deposits that have been mapped, measured, and economically recoverable are known as

   (A) ultimate reserves
   (B) sustainable reserves
   (C) proven reserves
   (D) clean reserves
   (E) known reserves

7. The force needed to accelerate 1 kilogram by 1 meter per second is called a

   (A) ton
   (B) meter per second
   (C) joule
   (D) calorie
   (E) newton

8. In the industrialized world, the burning of fossil fuels is probably the biggest single source of

   (A) air pollution
   (B) water pollution
   (C) urban blight
   (D) economic income of developed countries
   (E) international tensions between nations

9. Twenty percent of the total U.S. energy demand is used for lighting, heating, air conditioning, water heating, and

   (A) cooking
   (B) plastics
   (C) transportation
   (D) international export
   (E) construction

10. The first oil well in North America was drilled in Oil Springs, Ontario, in

    (A) 1855
    (B) 1875
    (C) 1905
    (D) 1935
    (E) 1955

11. Natural gas has much less waste since it needs little refining, is transported through underground pipelines, and burns with

    (A) boron as an additive
    (B) 75–95% efficiency
    (C) atmospheric particulates
    (D) smaller smoke stacks
    (E) the same efficiency as oil

12. Coal dust is very toxic to the lungs and causes

(A) arthritis
(B) cancer
(C) premature births
(D) black lung disease
(E) bad breath

13. Before fossil fuels became commonly used, what was popularly used for lamp oils and candles?

(A) Lye
(B) Bacon grease
(C) Whale oil
(D) Seal fat
(E) Natural gas

14. A sedimentary rock containing solid organic matter is called

(A) tephra
(B) kerogen
(C) granite
(D) igneous rock
(E) gneiss

15. Prior to mine reclamation laws, mine drainage and acidic runoff from coal tailings were toxic by-products of

(A) tillage
(B) urban development
(C) underground mining
(D) strip mining
(E) industrial processing

16. A hotly debated area with large oil reserves is

(A) NAWR
(B) USDA
(C) western Europe
(D) WARN
(E) ANWR

17. Coal contaminants must be removed by washing or flue-gas scrubbing or they will be released during burning to form sulfur dioxide (SO_2), sulfate (SO_4), CO_2, NO_x, and

(A) $C_8H_8O_3$
(B) particulates
(C) KCl
(D) CH_3NH_2
(E) Ga_2O_3

18. When coal is used to produce electricity, what fraction of the original energy is lost in thermal conversion at the plant?

(A) $\frac{1}{4}$
(B) $\frac{1}{3}$
(C) $\frac{1}{2}$
(D) $\frac{2}{3}$
(E) $\frac{3}{4}$

❯ Answers and Explanations

1. E

2. C—Sulfur is a big contaminant of coal when burned, and combines with oxygen into atmospheric pollutants.

3. D

4. E

5. B

6. C—Energy reserves must be proven or they have little value in energy planning.

7. E

8. A—Fossil fuel burning releases many different chemicals into the atmosphere.

9. A—Transportation and other U.S. energy demands far exceed cooking needs.

10. A—The first North American oil well was drilled over 150 years ago.

11. B—Natural gas is a clean-burning, high-efficiency fuel.

12. D—Black lung disease is an occupational hazard of coal mining, which causes inflammation and scarring of the lungs due to the long-term breathing of coal dust.

13. C—Whale oil was plentiful during the early years of the whaling industry and was used for lighting.

14. B—Kerogen is a fine-grain sedimentary rock also called oil shale.

15. D—Strip mining shears away rock and soil to get to deeper resource-bearing layers leaving huge amounts of waste rock, polluted water, and openly erosive surfaces.

16. E—The Alaska Arctic National Wildlife Refuge (ANWR) is home to thousands of caribou as well as millions of other species that ecologists fear will lose critical habitat during energy exploration.

17. B—Chemical pollutants, soot, and other particulates are released during coal burning.

18. D—When coal is burned, around 65% of its energy is lost at the plant during thermal conversion to electricity.

Free-Response Questions

1. The People's Republic of China leads the world in reliance on coal as a fuel source. Coal accounts for around 71% of China's total energy production and has fueled the country's swift rise in the global market. China's ever-increasing industrialization continues to increase the county's dependency on coal as its main fuel source. Though demand is currently outpacing production, China has enough coal reserves to sustain its economic growth for another century.

(a) What are the benefits of using coal over other fossil fuels?
(b) What problems might China face in its reliance on coal as a fuel source?
(c) What steps can China take to minimize the social and environmental risks of coal use?

2. The second law of thermodynamics states that the universe tends toward entropy (e.g., randomness). In terms of energy use, this means a fraction of an original energy amount is always lost after changing to another form. For example, much energy is lost through heat from turbines in a hydroelectric power plant.

(a) Can lost electric energy be retained?
(b) What modern materials might make energy transmission more efficient?

Free-Response Answers and Explanations

1.

a. The main reason coal provides a much better fuel source than other fossil fuels like oil and gas is because there is so much more of it. Global coal deposits are about 10 times greater than conventional oil and gas resources added together. Further, because it has been mapped and found economically recoverable, it is a proven resource. Though also a limited resource, it is extremely abundant.

b. Coal production for electricity can be extremely harmful to both the environment and public health. The process, from mining to waste disposal, creates both public and environmental risks. Because of hazardous substances contained within it (like mercury, sulfur dioxide, nitrogen oxides, and particulates), when coal is burned, it contaminates the air, land, and water.

c. Clean coal technology exists to help stem the tide of problems associated with using coal to produce electricity. Gasification can be used to convert coal into carbon monoxide and hydrogen, which can then be used as a fuel called synthetic gas. Chemicals can be used to wash the minerals and impurities out of coal, and flue gases can be treated with steam to prevent sulfur dioxide from escaping into the environment. Beyond clean coal technology, China can use alternative fuel sources like solar and wind power to help offset the current socioeconomic and environmental costs of coal use.

2.

a. Energy can be retained by keeping the amount and number of losses to a minimum.

b. Engineers and scientists are excited about the capabilities of carbon nanotubes as electrical transmission conduits. Individual carbon nanotube fibers have an electrical conductivity better than copper at only one-sixth the weight and with negligible current loss. Several researchers have demonstrated that a single-walled carbon nanotube can carry currents up to 20 microamperes. With current technology, losses in power transmission lines are about 7%. Dropping these losses to 6% would reap a national annual energy savings of 4×10^{10} kilowatt-hours (i.e., equal to about 24 million barrels of oil).

› Rapid Review

- The first law of thermodynamics explains that energy can be neither created nor destroyed.
- The second law of thermodynamics describes how the universe tends toward entropy (chaos, disorder, or randomness).
- In 1956, Shell Oil geophysicist M. King Hubbert calculated that the oil well extraction rate in the United States (lower 48 states) would peak around 1970.
- Fossil fuels are solids, liquids, and gases created through the compression of ancient organic plant and animal material in the Earth's crust.
- Currently, oil production in the United States from all sources is around 8 million barrels a day from over a half million wells.
- Fossil fuels produce around 85% of U.S. energy, with oil making up about 40% of that.
- Fossil fuel burning is the biggest single source of human-created air pollution in the industrialized world.
- Getting additional oil from a drilled deposit is known as secondary recovery.
- Canada is the biggest supplier of oil to the United States.

- In 1839, Abraham Gesner, a governmental geologist in Nova Scotia, discovered albertite, a coal-like material. Later he discovered a process to manufacture kerosene.
- Kerosene, or "coal oil," was cheap and smelled better than animal fat when burned.
- When coal is used to produce electricity, nearly 65% of the original energy is lost in thermal conversion at the plant.
- One calorie unit equals the amount of energy needed to heat 1 gram of water to 1 degree Celsius.
- The *International Energy Outlook 2008* projects that global energy consumption will increase by 50% from 2005 to 2030.
- Coal is often contaminated with sulfur, sometimes as much as 10% by weight.
- The *International Energy Outlook 2008* estimates that coal will account for 29% of total world energy consumption in 2030.
- Coal burning creates a lot of atmospheric pollution and particulates, and increasing its use will worsen the global greenhouse problem.

CHAPTER 16

Nuclear Energy

IN THIS CHAPTER

Summary: By the mid-1990's, roughly 20% of America's electricity was provided by licensed power reactors. Nuclear energy, using ^{235}U, has the capability to provide energy for several hundred years.

Keywords

- Fission, fusion, breeder reactor, uranium, Three Mile Island, radioactivity, Chernobyl, Nuclear Regulatory Commission, half-life, radiation

Nuclear Energy

Currently, nuclear energy provides the world with around 15% of its energy needs. In the United States, 20% of electricity comes from nuclear power generating plants. In fact, Vermont leads the nation with 85% of its electricity from nuclear power.

Public perception fluctuates with regard to nuclear power. When everything is running smoothly and providing clean energy, most people are happy. However, accidents happen, and when they do, nuclear energy claims more than its share of bad public opinion, where radioactive health risks and radioactive waste are show stoppers.

Uranium

Uranium is commonly found in minute amounts of the uranium oxide mineral *uraninite* (*pitchblende*) in granite and other volcanic rocks. Natural uranium is 99.3% ^{238}U and 0.7% ^{235}U. Nuclear energy comes primarily from uranium ^{238}U enriched with 3% ^{235}U.

Fission and Fusion

Nuclear power is created in one of two ways; *fission* (splitting) of uranium, plutonium, or thorium atoms; and *fusion* (merging) of two smaller atoms into one larger atom with a combined nucleus. As the nuclei merge, energy is released, but this method is scientifically complex and has never been shown to produce more energy than it consumes.

Nuclear Reactors

Most *nuclear reactors* contain a core with a large number of fuel rods loaded with uranium oxide pellets. The fuel pellets (usually around 1 cm diameter and 1.5 cm long) are commonly arranged in a long zirconium alloy tube to form a fuel rod—the zirconium is hard, corrosion resistant, and permeable to neutrons. Up to 264 control rods of neutron-absorbing material (e.g., silver, cadmium, boron, or hafnium) can be inserted or lifted from the core to control the reaction rate or to stop it. A moderator material, like water or graphite, slows down neutrons released from fission so more fission is created.

When a ^{235}U isotope undergoes fission, it absorbs a neutron and then splits into two fission pieces (and other atomic particles) that ricochet away at high velocity. When they stop, their energy is converted to heat—about 10 million times the heat of burning a carbon atom in coal.

Boiling water and *pressurized water reactors* are two types of similar nuclear reactors. They both have a hot reactor core, which heats water to steam and, in turn, spins turbines to generate electricity. Pressurized reactors use heat from the core to heat a second water supply for the turbines via a heat exchanger, and a third water system to cool steam from the turbines to be used again.

Breeder Reactors

Under the right operating conditions, neutrons given off by fission reactions can "breed" or create more fuel from otherwise nonfissionable isotopes. *Breeder reactors,* which produce more energy than they use, can provide energy for billions of years. The most common breeder reaction is that of plutonium (^{239}Pu), a by-product of nonfissionable ^{238}U. Uranium's most stable isotope, ^{238}U, has a half-life of nearly $4\frac{1}{2}$ million years. However, since breeder reactors use uranium by-products, which are extremely dangerous and can be used in nuclear weapons, policy makers are cautious about their construction.

Nuclear Capacity

Currently, there are 436 nuclear power plants worldwide, with 104 of those in the United States (as of June 2009). There are another 48 plants under construction in 15 countries. Figure 16.1 illustrates the parts of a nuclear power plant.

Water, heated to steam, drives a turbine which generates electricity. These plants produce nearly 20% of the world's total electricity. In addition, they create huge amounts of reliable energy from small amounts of fuel without the air pollution created by burning fossil fuels. However, reactors are only licensed for 40 years, and early reactors in the United States have been or are in the process of being decommissioned.

In 1973, only 83 billion kilowatt-hours (kWh) of nuclear power was produced. In 2005, nuclear power produced 2.6 *trillion* kWh of electricity and is expected to approach 4 trillion kWh by 2030. Despite the initial growth in U.S. nuclear power, the threat of radioactive release, waste disposal, and/or terrorism put a stop to it. In fact, no new nuclear power plants have been built since 1977.

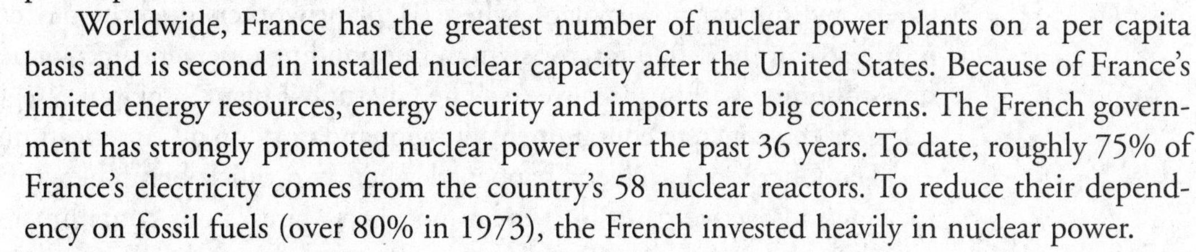

Worldwide, France has the greatest number of nuclear power plants on a per capita basis and is second in installed nuclear capacity after the United States. Because of France's limited energy resources, energy security and imports are big concerns. The French government has strongly promoted nuclear power over the past 36 years. To date, roughly 75% of France's electricity comes from the country's 58 nuclear reactors. To reduce their dependency on fossil fuels (over 80% in 1973), the French invested heavily in nuclear power.

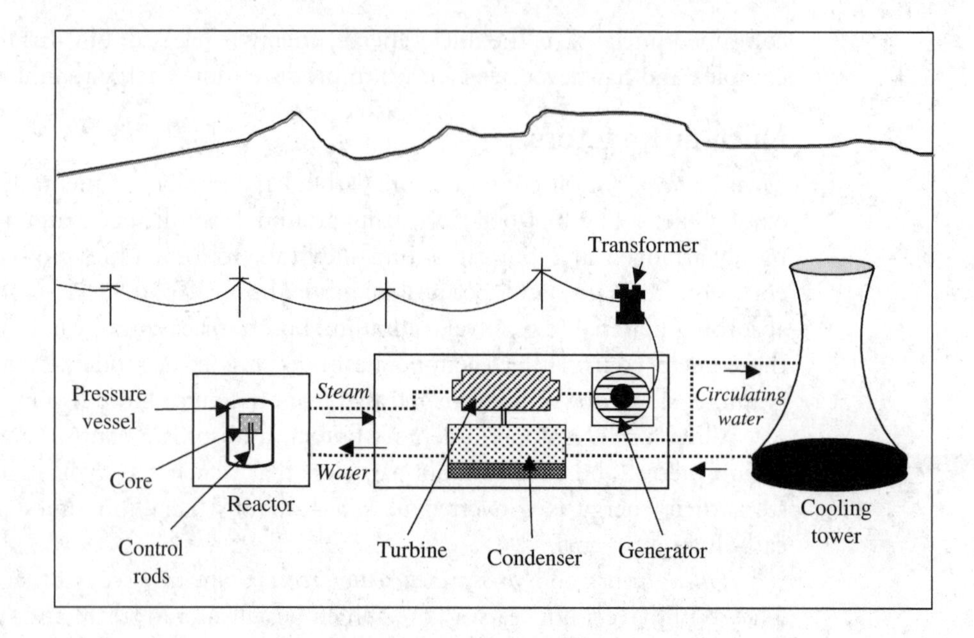

Figure 16.1 An impoundment hydroelectric power plant is commonly known as a dam.

Accidents

Nuclear power is reliable, but safety is critical since a nuclear accident can cause a huge environmental disaster.

Three Mile Island

On March 28, 1979, the nuclear accident at *Three Mile Island,* a two-unit nuclear plant on the Susquehanna River in Pennsylvania, caused great safety concerns and fears of radiation leakage.

At Three Mile Island, the reactor lost cooling water and overheated, and some of the fuel rods melted and ruptured. This resulted in a release of radioactive gases into the atmosphere and critical damage to the reactor. People within a 1-mile radius of the reactor were evacuated, but no one was injured.

The U.S. nuclear industry has been stopped in its tracks since the Three Mile Island accident, the worst nuclear accident in U.S. history. No company has continued with plans to build a new nuclear plant since the accident, even though former President George W. Bush backed new construction of nuclear plants as part of his energy policy.

In a possible public opinion change, some environmentalists are starting to rethink nuclear power, since unlike fossil fuels, it doesn't produce enhanced greenhouse gas emissions. Modern nuclear plant designs also contain smaller reactors, which create less radioactive waste and take into account 75 factors including human factors, seismic activity, water availability, and emergency preparedness issues.

Chernobyl

The world's worst nuclear plant accident occurred in the Ukraine, on April 26, 1986, at the Chernobyl nuclear power plant, where 31 plant workers died the day of the accident and nearly 200 of the 1,000 reactor staff and responding emergency personnel died from radiation poisoning within three weeks. The Chernobyl plant, a graphite-regulated reactor, did not have the concrete containment dome mandatory on all American nuclear plants.

The Chernobyl accident happened when two quick blasts blew off the reactor roof, spewing radioactive gases into the atmosphere. Fires, poor containment design, operator mistakes, and a power surge led to the catastrophic and deadly conditions. With the core continuing to heat up, Russian officials ordered aircraft to dump 5,000 tons of lead, sand, and clay onto the site to bring the temperature down.

The East-West politics of the time resulted in an early cover-up of the incident, until Swedish scientists detected the radioactive cloud moving across several countries and demanded an explanation.

Immediately after the Chernobyl accident, 116,000 people were relocated from the contaminated area. Later, another 350,000 were relocated from the most severely contaminated areas. Health studies linked increased birth defects and 4,000 thyroid cancer cases in children living in the Chernobyl area at the time to the accident.

Although radiation has nearly returned to baseline levels in much of the surrounding area, inhabitants were traumatized by relocation, economic losses, and the need for young people to seek jobs elsewhere. The number of deaths still tops births and gives the further perception that the area is tainted.

The China Syndrome

In addition to radioactivity, the spiraling temperature rise during a meltdown is a huge safety issue. This is known as the China Syndrome.

> The *China Syndrome* is a hypothetical concept that refers to what might happen if nuclear core temperatures escalated out of control, causing the core to melt through the Earth's crust all the way to the other side of the world (to China).

In 1979, a Hollywood movie entitled *The China Syndrome* described incidents that occurred at the Rancho Seco power plant in California. However, unlike the movie portrayal, the Rancho Seco emergency system shut the reactor down safely as it was designed to do.

Japan's Earthquake/Tsunami

The Japanese Kashiwazaki-Kariwa plant is the world's largest nuclear power plant in power-output with seven reactors generating 8.2 million kilowatts of electricity. It is located 135 miles northwest of Tokyo. On March 11, 2011, an 8.9 magnitude earthquake shook the coast of Japan and created a 10-meter (33-ft) tsunami. The earthquake-damaged and swamped nuclear plant released radioactive coolant water when pipes cracked and a containment building exploded. Contaminated radioactive water leaked into the Sea of Japan and ten days later, milk and spinach were found to have dangerous, elevated reactor byproducts (e.g., radioactive iodine-131 and cesium-137) within 105 km (65 mi) of the plant. High levels of iodine-131 are known to cause thyroid cancer, while cesium-137 damages cells and can cause cancer. As of August 2011, the official death toll reported by the Japan Fire Department from the earthquake and tsunami was 21,234 (16,477 deaths and 4,787 missing). They also reported 111,944 buildings destroyed with over 600,000 additional buildings partially destroyed and/or damaged.

Radioactivity

French scientist Marie Curie used the term *radioactivity* for the first time in 1898. Curie and her physicist husband Pierre found that radioactive particles were emitted as either electrically negative (–) *beta* (β) *particles* or positive (+) *alpha* (α) *particles.*

Radioactivity is considered a bad side effect of nuclear weapons and x-rays, but when properly shielded, radioactive elements are useful. In fact, radioactive elements provide power sources for pacemakers, satellites, and submarines.

Nuclear Reactions

Most chemical reactions are focused on the outer electrons of an element, sharing, swapping, and bumping electrons into and out of the combining reaction partners. However, nuclear reactions are different. They take place inside the nucleus.

TIP

There are two types of nuclear reactions. The first is the radioactive decay of bonds within the nucleus that emit radiation when broken. The second is the "billiard ball" type of reaction where the nucleus or nuclear particle (like a proton) is struck by another nucleus or nuclear particle.

Radioactive Decay

A radioactive element, like everything else in life, decays (ages). When uranium or plutonium decays over billions of years, it goes through a transformation process of degrading into lower-energy element forms until it settles into one that is stable.

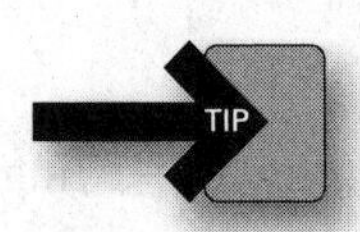

When a radioactive element decays, different nuclear particles are given off. These radiation particles can be separated by an electric (magnetic) field and detected in a laboratory as

Beta (β) *particles* = negatively (–) charged particles

Alpha (α) *particles* = positively (+) charged particles

Gamma (γ) *particles* = electromagnetic radiation with no overall charge (similar to x-rays), but with a shorter wave length

Radioactive isotope decay is affected by an element's stability at a certain energy level. An instrument used to detect radioactive ions is called an *ionization counter.* Bismuth (Bi), atomic number 83, is the heaviest element in the periodic table with one stable isotope. Other heavier elements (e.g., thorium, einsteinium) are radioactive.

Half-life

All radioactive isotopes have a specific *half-life.* These are not dependent on pressure, temperature, or bonding properties, but on the specific energy levels of the isotope's molecular makeup.

> The *half-life* of a radioactive isotope is the time it takes for one-half of an elemental sample to decay.

For example, the half-life of plutonium ^{239}P is 2.13×10^6 years. The half-life of ^{238}U is 4.5×10^9 years, about the same age as the Earth. It is sobering to think that the uranium found today will be around for another 4 billion years.

Nuclear Disadvantages

A serious problem with nuclear power is the storage of radioactive waste. Each year, about 30 metric tons of used fuel are created by every 1,000-megawatt (MW) nuclear electric power plant. Most of this waste is stored at the power plants because of the lack of high-level radioactive waste disposal sites. Long-term storage is crucial as additional radioactive waste accumulates.

Although not much waste is created at any one plant, it is extremely dangerous. It must be sealed up and buried for decades, even centuries, to allow time for the radioactivity to gradually disappear.

Radioactive Waste Storage

In January 2002, Energy Secretary Spencer Abraham recommended Yucca Mountain, Nevada, as the nation's permanent nuclear waste depository. His plan called for specially designed containers that would hold over 77,000 tons of nuclear waste from both power plants and nuclear weapons' manufacturing to be buried in a series of tunnels dug 302 meters below the mountain's peak and 274 meters above the water table.

Work has been halted on the Yucca Mountain site and it no longer receives federal funding. In 2012, President Obama's Blue Ribbon Commission on America's Nuclear Future submitted its final report to the Secretary of Energy on potential alternatives for nuclear waste storage.

When Congress members began talking about reviving the Yucca Mountain Project in September 2014, Senate Majority Leader Harry Reid (D-NV) who is strongly opposed to the Yucca Mountain nuclear storage facility in his home state told reporters, "Yucca Mountain is all through. As long as I'm around, there's no Yucca Mountain."

Geologists are studying the pros and cons of storing radioactive waste beneath the deep oceans and in stable (no tectonic activity) rock formations, while the U.S. Department of Energy is again looking at trains to haul 150-ton casks filled with used, radioactive nuclear fuel waste from nuclear power plants to disposal sites that haven't been built yet.

Now the mothballed Nevada storage facility that the U.S. government spent $15 billion to construct is dormant while thousands of tons of spent nuclear fuel build up at nuclear reactors around the country. The big concern is that groundwater can seep into land sites and become contaminated. States like Nevada, designated as radioactive repositories, are unhappy about the significant safety hazards. Since the earth continues to surprise geologists, there is no way to guarantee a chosen site will be stable and safe for radioactive storage for hundreds or thousands of years. The radioactive storage debate will continue for a long time.

Transporting Waste

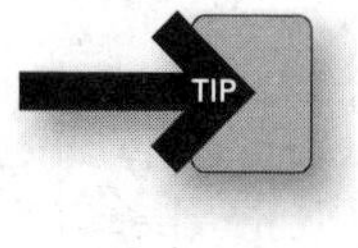

Another big controversy centers on the transportation of nuclear waste to any distant storage site by rail or truck. Since most nuclear power plants are in the eastern part of the United States, the waste must be transported about 2,000 miles westward if relatively unpopulated regions are to be considered for long term waste storage.

Besides the usual transportation mishaps possible on a long trip, the very real threat of terrorist sabotage hangs like a black cloud over the entire transportation plan. In any case, nuclear utilities are running out of radioactive waste storage capacity. If we continue to use nuclear power, we'll have to store the waste somewhere.

Regulations

Compared to the extreme Chernobyl accident, the Three Mile Island incident was not serious and no deaths occurred. In fact, the best thing to come out of Three Mile Island was much better plant and safety designs and the creation of the *Institute of Nuclear Power Operations* (INPO). The INPO established guidelines for excellence in nuclear plant operations and increased communication within the nuclear industry.

A strong regulatory impact in response to public outrage has been a big factor in the growth of nuclear power. In 2000, the *Nuclear Regulatory Commission* (NRC), in an encouraging nod to the U.S. nuclear power industry, granted the first-ever renewal of a nuclear power plant's operating license. The 20-year extension (2034 and 2036 for two reactors) was given to the 1700-MW Calvert Cliffs, Maryland, plant.

A 2011 U.S. opinion poll has shown a turn around with 62% of those polled favoring nuclear power production compared to 35% opposed. In fact, Southern Company (a utility company) has begun construction on two new nuclear units that are expected to provide commercial power by 2016 and 2017, respectively.

› Review Questions

Multiple-Choice Questions

1. A big concern facing geologists in long-term radioactive waste storage is that

(A) people living in the area will begin to glow in the dark
(B) the half-life will take even longer
(C) groundwater will seep into land sites and become contaminated
(D) people will mistake it for fluorescent minerals
(E) the public will not tolerate the necessary safety measures

2. The isotope of uranium that undergoes fission and releases huge amounts of energy is

(A) ^{190}U
(B) ^{225}Pt
(C) ^{60}I
(D) ^{235}U
(E) ^{90}Ba

3. A nuclear reactor contains a core with many fuel rods containing

(A) steel pellets
(B) uranium oxide pellets
(C) iron pellets
(D) palladium pellets
(E) silicon pellets

4. Which country's government has strongly promoted increases in nuclear power use over the past 30 years?

(A) Columbia
(B) Madagascar
(C) Holland
(D) France
(E) Spain

5. An instrument used to detect radioactive ions is called a(n)

(A) mass spectrometer
(B) metal detector
(C) ionization counter
(D) gas chromatograph
(E) high-performance liquid chromatograph

6. A strong regulatory backlash to nuclear power in the United States occurred in response to the

(A) Three Mile Island accident
(B) first U.S. nuclear power plant built
(C) September 11, 2001, terrorist attacks
(D) eruption of Mount St. Helens
(E) use of nuclear power in submarines

7. No new nuclear power plants have been built or planned in the United States since

(A) 1958
(B) 1962
(C) 1977
(D) 1983
(E) 1990

8. The time for a radioactive element or isotope to degrade is known as its

(A) half-life
(B) fission potential
(C) transmutability point
(D) radioactive potential
(E) radioactive point

9. The Yucca Mountain, NV, facility, originally scheduled to open in 2017, was intended for the

(A) mining of uranium
(B) Department of Energy's yearly retreat
(C) storage of new power plant construction material
(D) protection of local inhabitants in the event of nuclear attack
(E) storage of spent uranium fuel

10. The biggest advantage of nuclear energy generation is that

(A) it is free
(B) it doesn't generate greenhouse gases
(C) it takes very few people to operate a plant
(D) it produces low-wattage power
(E) many developing countries are rapidly increasing their nuclear portfolio

11. Decay of radioactive isotopes is affected by the

(A) isotopes' boiling point
(B) pressure
(C) properties of the isotopes' storage container
(D) stability of an element at a certain energy level
(E) size of the original sample

12. After Three Mile Island, much better plant and safety designs, along with the creation of which organization, occurred?

(A) Department of Energy
(B) Occupational Health and Safety Organization
(C) Department of the Interior
(D) Richard Smalley Nanotechnology Institute
(E) Institute of Nuclear Power Operations

13. Each year, roughly what amount of used fuel is created by every 1,000-megawatt nuclear electric power plant?

(A) 10 metric tons
(B) 25 metric tons
(C) 30 metric tons
(D) 50 metric tons
(E) 65 metric tons

14. French scientist Marie Curie coined what term in 1898?

(A) Radioactivity
(B) Isotope
(C) Half-life
(D) Geological time
(E) Fission

15. Besides alpha and beta particles separated in a magnetic field, what other short-wavelength particles can be detected?

(A) Joules
(B) Quarks
(C) Gamma particles
(D) Radioactive dust
(E) Omega particles

16. Radioactive elements provide power sources for all but which of the following?

(A) Pacemakers
(B) Generators
(C) Submarines
(D) Satellites
(E) Automobiles

17. What important action did the Nuclear Regulatory Commission take in 2000?

(A) They decided to stop work on the Yucca Mountain waste storage site.
(B) They issued the first-ever renewal of a nuclear power plant's operating license.
(C) They offered green credits for nuclear energy.
(D) They published a successful method of nuclear fusion.
(E) They collaborated with France on better ways to store electric power.

18. Nuclear power is commonly generated using the element

(A) copper
(B) actinium
(C) mercury
(D) uranium
(E) platinum

› Answers and Explanations

1. C—States designated as radioactive repositories are concerned about safety hazards.

2. D—Nuclear energy comes from enriched uranium (^{235}U).

3. B—The fission process uses uranium oxide pellets.

4. D

5. C—When uranium undergoes fission, high-energy ions are given off.

6. A

7. C

8. A—A radioactive isotope's half-life is the time it takes for one-half of an elemental sample to decay.

9. E—In specially designed containers, over 77,000 tons of nuclear waste from power plants and nuclear weapons will be buried in deep underground storage tunnels.

10. B—If you don't count radioactive waste, nuclear energy is nonpolluting to the air.

11. D—Bismuth is a stable, heavy, non-radioactive element.

12. E—The INPO established guidelines for excellence in nuclear plant operations and increased communication within the nuclear industry.

13. C

14. A—It was used to describe the energy generated by certain elements.

15. C—Gamma (γ) particles are electromagnetic radiation with no overall charge, but a shorter wavelength.

16. E—Currently, no automobiles have been designed to run on nuclear power.

17. B—The 20-year extension was issued for two reactors at a Maryland plant.

18. D

Free-Response Questions

1. France appears to be rethinking its nuclear push. Even though the French government planned to eventually have 100% nuclear power generation, strong environmental concern slowed nuclear growth. When Germany chose to phase out nuclear power, French public concern grew. Currently, French opinion polls favor an end to nuclear power.

Unfortunately, the French have no clean, affordable substitutes to handle the power demand. They must either replace aging nuclear plants with modern ones, or phase out nuclear power, since several reactors need replacement before 2020.

(a) Describe the difficulties France may encounter when attempting to switch from nuclear energy to lesser developed energy generators.
(b) What are the technological alternatives?
(c) Describe how these various alternatives might meet the demand.
(d) How might the switch be speeded up?

2. After discovery of corrosion in a major nuclear plant section in Ohio, the Nuclear Regulatory Commission ordered safety information on 68 other units. After checking the problem thoroughly, the problem was found to affect only the Ohio unit. The news media reported this problem thoroughly when it was discovered, but didn't follow-up on the other inspected units months later.

(a) Describe the impact of the initial story on the public's perception of nuclear safety.
(b) What should the government have done to allay public concern about U.S. nuclear energy and its safety record?
(c) Did this incident promote the argument for the construction of new nuclear plants?

Free-Response Answers and Explanations

1.

a. There are several socioeconomic and environmental problems France may face. Added to shifting public opinion, the costs of completely changing energy-producing technologies would create a huge economic strain on France's government and population. Plus, energy prices would increase considerably, placing even more strain on the population. Disposing of hazardous nuclear waste is also a concern.

b. Aside from various fossil fuel sources (e.g., coal, oil, and natural gas), there are clean energy sources available. Solar and wind power provide inexhaustible sources of energy whose costs are steadily declining thanks to market forces and government subsidies.

c. Alternatives such as solar and wind power would have to be used together and/or in conjunction with other energy sources, like fossil fuels, to meet the demand.

d. The switch could be sped up by beginning to phase out nuclear energy with other sources like wind and/or solar power in specific communities. It could also be accelerated if communities took more ownership of their own energy power needs.

2.

a. Since the general public knows so little about nuclear plants, the story's initial impact would have caused alarm. Public fears about nuclear energy sources, especially after catastrophic events in the not too distant past due to plant failures, are easily aroused and can lead to panic.

b. To allay public concern about the safety of nuclear energy, the government should have fully reported on the 68 safe units. The government must do a much better job of informing the public about nuclear energy. With more information, public opinion would be better informed and less likely to be aroused in a similar event.

c. Yes and no. For those whom the story aroused fear about the prospect of nuclear energy, this incident would have prompted an argument for shutting down nuclear power plants and looking for alternative energy sources. However, those informed of nuclear energy's safety and benefits might have been persuaded by the incident to build new plants or update those in need of retooling.

❯ Rapid Review

- Uranium's most stable isotope, ^{238}U, has a half-life of nearly $4\frac{1}{2}$ billion years.
- Uranium oxide, found in granite and other volcanic rocks, is known as pitchblende.
- Nuclear power is created by fission (splitting) of uranium, plutonium, or thorium atoms; and fusion (merging) of two smaller atoms into a larger atom with a combined nucleus.
- Nuclear energy, using ^{235}U, has the capability to provide energy for several hundred years.
- Most nuclear reactors contain a core with a large number of fuel rods loaded with uranium oxide pellets.
- Breeder reactors produce more energy than they use.
- The most common breeder reaction is that of plutonium (^{239}Pu), a by-product of nonfissionable ^{238}U.
- Fires, poor containment design, operator mistakes, and a power surge led to the catastrophic accident in Chernobyl, Ukraine.
- The Chernobyl plant, a graphite-regulated reactor, did not have the concrete containment dome mandatory on all American nuclear plants.

- The China Syndrome is a hypothetical concept that refers to what might happen if nuclear core temperatures escalated out of control, causing the core to melt through Earth's crust all the way to the other side of the world (to China).
- Boiling water and pressurized water reactors are two types of similar nuclear reactors with hot reactor cores that heat water to steam, spin turbines, and generate electricity.
- The INPO established guidelines for excellence in nuclear plant operations and increased communication within the nuclear industry.
- The Curies found that radioactive particles were emitted as electrically negative beta (β) particles or positive alpha (α) particles.
- At Three Mile Island, the reactor lost cooling water, overheated, and the fuel rods melted and ruptured; in addition radioactive gases were released into the atmosphere.
- When a radioactive element decays, different nuclear particles (e.g., alpha, beta, and gamma) can be separated by a magnetic field.
- Gamma (γ) particles have no overall charge, but a shorter wavelength.
- Transportation of nuclear waste by rail or truck is an issue because of possible radiation release from rail or highway accidents and terrorism.

Alternative and Renewable Energies

IN THIS CHAPTER

Summary: Global dependence on fossil fuels is resource limiting. Other non-polluting, renewable, and efficient sources of energy must be developed. Possible options include hydroelectric power, solar energy, hydrogen fuel cells, tidal, and biomass energy resources.

Keywords

✪ Clean fuels, ethanol, photovoltaic, semiconductor, Staebler-Wronski effect, distributed power, polycrystalline materials, electrochemical solar cells, geothermal, tidal energy

---

Clean Fuels

The most common fuels used for transportation in the United States are gasoline and diesel fuel, but there are several additional energy sources able to power motor vehicles. These include alcohols, electricity, natural gas, and propane. When vehicle fuels, because of physical or chemical properties, create less pollution than gasoline, they are known as *clean fuels.*

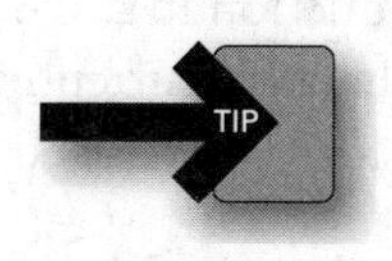

Clean energy is produced from renewable energy sources (e.g., solar, wind, biomass, hydroelectric, and geothermal). Figure 17.1 shows world energy use from various sources.

Electricity

Battery-powered or *hybrid gas-electric vehicles* are a great option for a lot of commuters. A hybrid vehicle, such as the Toyota Prius, has a 1.5-liter (L) gas engine that only kicks in when the car is accelerating or going uphill, or when the battery needs recharging. The Prius is

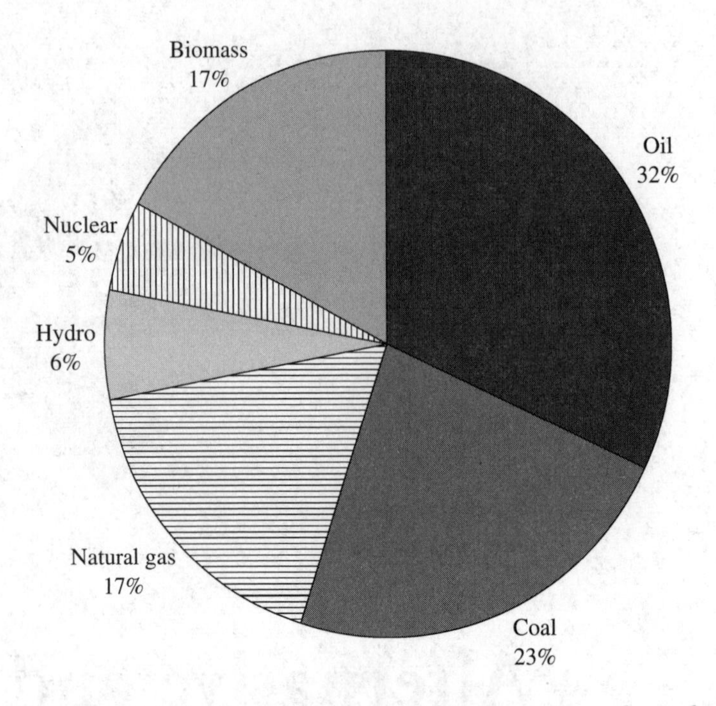

Figure 17.1 Over 70% of world energy choices originate from fossil fuels.

rated at 60 miles per gallon (mi/gal) (25 km/L) in city driving and 51 mi/gal (22 km/L) on the highway. Practically pollution free, it offers a great option for lowering vehicle emissions in polluted cities.

Low-power batteries limit electric cars' driving range, but this is improving. With batteries taking minutes instead of hours to recharge and running longer distances on one charge, electric power could become a widespread clean fuel for the future.

Biomass

Plants use sunlight for energy. This energy is released as heat when plants (*biomass*) are burned. Forty percent of the world's population uses wood or charcoal as a main energy source, with the poorest countries (e.g., Ethiopia and Burundi) using biomass for 90% of their energy.

Biomass and *biofuels* are also produced from corn and other crops, as well as from wood or paper wastes. Since these renewable resources pull carbon dioxide out of the atmosphere as they grow, they decrease greenhouse gas buildup.

Ethanol

Ethanol (grain alcohol) is created by anaerobic digestion of high-sugar-content plants (e.g., grain and sugar cane). Ethanol-gasoline blends, known as *gasohol,* have been used in the United States and Brazil for many years. Pure ethanol gives great engine performance along with low hydrocarbon and toxic emissions.

Forty-two percent of the U.S. corn crop was used for ethanol fuel production in 2014. It is also added to gasoline to reduce incomplete combustion emissions. Ethanol is currently more expensive than gasoline, but as fossil fuels run out and new processing is developed, prices will improve.

The disadvantage of ethanol is its ability to augment the polluting effects of some compounds, like benzene. Ethanol acts as a solvent and slows the breakdown of benzene, toluene, and other chemicals in soil and groundwater. The longer these highly toxic compounds stay in the soil, the greater the public health risk.

Methanol

Methanol (wood alcohol), like ethanol, is a high-performance liquid fuel that releases low levels of toxic and ozone-forming compounds. It can be made for about the same cost as gasoline from natural gas, wood, and coal. All major auto manufacturers have produced cars that run on M85, a blend of 85% methanol and 15% gasoline. Cars burning pure methanol (M100) offer much greater air quality and efficiency advantages. Race cars use methanol because of its superior performance and fire safety characteristics.

Solar Energy

Sunlight is an immense source of natural energy. It lights and heats our planet and supplies energy for plant photosynthesis. The amount of solar energy reaching the Earth annually is much greater than worldwide energy demands, although it fluctuates with the time of day, location, and season.

Solar technologies use the sun's energy and light to provide heat, light, hot water, electricity, and cooling for homes, businesses, and industry. Accessibility to unblocked sunlight for use in both passive solar designs and active systems is protected by zoning laws and ordinances in many parts of the country.

Crystalline Silicon Solar Cells

Most solar power is based on the *photovoltaic* (PV) reaction, which produces voltage when exposed to radiant energy (especially light). A solar PV cell converts sunlight into electricity. Crystalline silicon is the light-absorbing semiconductor developed in the microelectronics industry and used in many solar cells.

Two types of crystalline silicon are used. The first, monocrystalline silicon, is produced by slicing thin wafers (up to 0.150 cm diameter and 350 microns thick) from a single, high-purity crystal. The second, multicrystalline silicon, is made by cutting a cast block of silicon into bars, and then wafers. It is the most common substrate used by silicon cell manufacturers.

For both monocrystalline and multicrystalline silicon, a *semiconductor* connection is made by diffusing phosphorus onto the top of a boron-coated silicon wafer. Contacts are applied to the cell's front and back, with the front contact pattern designed to allow maximum light exposure to the silicon material. The most efficient solar cells use monocrystalline silicon with covered, laser-grooved, grid contacts for maximum light absorption and current.

Efficiency

Solar cells function more efficiently under focused light. Unlike common flat-plate PV arrays, concentrator systems need direct sunlight and don't operate in cloudy conditions. They follow the sun's path through the sky using *single-axis tracking*. To follow the sun's changing height seasonally, two-axis tracking is used. For this reason, mirrors and lenses are used to direct light onto specially designed cells with heat sinks or active cell cooling and to disperse the high heat created. An average crystalline silicon cell has an efficiency of 15%, compared to an average thin-film cell with an efficiency of around 6%.

Thin-Wafer Solar Cells

The high cost of crystalline silicon wafers led the semiconductor industry to look for cheaper ways and materials to make solar cells. The most commonly used materials are *amorphous silicon* or *polycrystalline materials* (e.g., cadmium telluride, gallium). These materials strongly absorb light and are about 1 micron thick, so production costs drop.

These thin wafers allow large area deposition (up to 1 meter) and high-volume manufacturing. Thin-film semiconductor layers are deposited onto coated glass or stainless-steel sheets. Intricate thin-film methods have taken about 20 years to get from initial research to prototype manufacturing.

Staebler-Wronski Effect

TIP

Amorphous silicon is the best thin-film technology with a single sequence of layers, but output drops (15–35% loss) when exposed to the sun. Not a good thing for *solar* cells! Solar cell degradation is called the *Staebler-Wronski effect.* It describes how the best stability uses the thinnest layers to increase the electric field strength across the material, but reduces light absorption and cell efficiency. The best thin-film cells are low cost, laminated for weather resistance, and have high stable efficiencies and yields.

Electrochemical Solar Cells

While crystalline and thin-film solar cells have solid-state light-absorbing layers, *electrochemical solar cells* use a dye sensitizer to absorb light and produce electron pairs in a nanocrystalline titanium dioxide semiconductor layer. Although cheaper, companies' ability to scale up electrochemical PV cell manufacturing will be proven over time.

Passive Solar

Buildings designed for *passive solar* and day lighting use design features like large south-facing windows and construction materials to absorb and slowly release the sun's heat. No mechanical processes are used in passive solar heating. Incorporating passive solar designs can reduce heating bills as much as 50%. Passive solar designs can also provide cooling through natural ventilation.

Off the Grid

More and more PV installations on homes and buildings are being connected to the electricity grid. Demand is encouraged by governmental programs (Japan and Germany) and incentive pricing and electricity providers (Switzerland and United States). The main push comes from individuals and/or companies who want to get electricity from a clean, nonpolluting, renewable source and agree to pay extra for the option.

An *electricity grid* is an electricity transmission and distribution system, usually supplying power across a wide geographical region.

An individual PV system, connected to a larger supply grid, can supply electricity to a home or building. Any extra electricity can be sent to the grid. Batteries are not needed since the grid is able to meet any extra demand. However, for a home to be independent of the grid, battery storage is needed for power at night.

Grid-connected systems are independent power systems joined to a regional grid that draw on the grid's reserve capacity when they need it and give electricity back during times of extra production.

One disadvantage of solar power is the amount of land needed. For enough solar panels to supply the electricity needs of an urban area, lots of acreage is needed.

Common Uses

Solar home systems are made up of a PV panel, rechargeable battery to store the energy captured during daylight hours, regulator, and necessary wiring and switches. Solar PV modules can be added to a pitched roof above the existing roof tiles, or tiles can be replaced by specially designed PV roof tiles or roof-tiling systems.

Cost-effective PV systems are also great for beach or vacation homes, or remote cabins without access to an electricity grid. Solar energy power is highly reliable and needs little maintenance, making it a great choice for remote sunny locations. Polar research stations are a good example.

Central power applications use solar energy in the same way a traditional utility company operates a major power station. There are hub locations from which power is sent out to meet demand.

Power sent out in small amounts, usually near the point of electrical usage, is known as *distributed power.*

Solar energy is used for industrial applications where only a few kilowatts of power are needed. These applications include powering microwave repeater stations, TV and radio, telemetry and radio telephones, as well as school traffic lights.

Solar power is also used for transportation signaling (navigation buoys, lighthouses, and airstrip warning lights). Environmental monitoring equipment, as well as pipeline corrosion safeguard systems, wellheads, bridges, and other structures also use solar power. Apart from off-grid homes, other remote buildings such as schools, community halls, and clinics all benefit from solar energy to power TV, video, telephone, and refrigeration equipment.

In some rural areas, solar panels are configured as central village power plants, which power homes through a wired network, or act as a battery-charging station where local people can recharge home batteries.

PV systems pump water in remote areas as part of portable water supply systems and desalination plants. Larger off-grid systems are constructed to power higher and more sophisticated electrical loads with an array of PV modules and more battery capacity.

To meet the largest power requirements in an off-grid location, a PV system is sometimes configured with a small diesel generator. This means the PV system no longer has to meet low-sunlight conditions. The diesel generator provides backup power, but is rarely used by the PV system, so fuel and maintenance costs stay low, and diesel use is minimal.

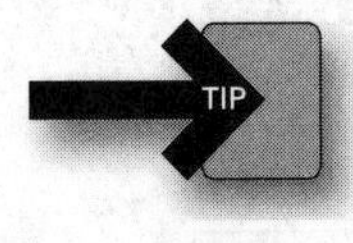

The bottom line is that the Earth gets more energy from the sun in an hour than the planet uses in a whole year. Since 2 billion of the world's people have no access to electricity, solar power is an excellent, renewable, and nonpolluting energy option.

Wind Energy

Wind energy is created by the Earth's atmospheric circulation patterns, which are heated and influenced by the sun. This *green energy* source creates electricity without consuming natural resources or producing greenhouse gases.

Wind power is converted into electricity through large, high-tech turbines built on a high tower to capture the greatest amount of wind. At 30+ meters above the ground, they can catch faster and less turbulent wind. Turbines catch the wind with two to three

Table 17.1 Wind turbines are relatively quiet compared to many modern noises.

SOUND	DECIBEL LEVEL (DB)
Rustling leaves	20
Whispering	25
Library	30
Refrigerator	45
Normal conversation	60
Wind turbine	60
Washing machine	65
Dishwasher	65
Car	70
Vacuum cleaner	70
Busy traffic	75
Alarm clock	80
Noisy restaurant	80
Outboard motor	80
Electric shaver	85
Screaming child	90
Passing motorcycle	90
Live rock music	90–130
Subway train	100
Diesel truck	100
Jackhammer	100
Helicopter	105
Lawn mower	105
Sandblasting	110
Auto horn	120
Airplane propeller	120
Air raid siren	130
Gunshot	140
Jet engine	140
Rocket launch	180

propeller-like blades mounted on a shaft to form a *rotor* that spins a generator, creating electricity. Electricity is sent through power lines to locations far from the turbines.

In the United States, coastal areas, midwestern plains, and mountain passes all funnel and raise wind speeds. In the first half of 2014, wind provided 5% of the nation's renewable-based electricity. With production tax credits, wind energy could provide 150 gigawatts or 20% of the nation's electricity by 2020.

After the United Nations' Kyoto accord in 1997, 160 industrialized nations committed to lowering average greenhouse emissions (5% below 1990 levels) by 2012, but several nations have since withdrawn support and the accord has stalled. To do this, many nations looked to wind power. Globally, wind energy capacity grew by 29% to reach global installations of 121 gigawatts (GW) and cover 3% of the world's total energy demand. Denmark, the largest user of wind energy, will obtain 35% of its total energy from renewable (half from wind farms) by 2020 and 100% by 2050.

Disadvantages of Wind

Drawbacks of wind power are unpredictability and the vast acreage needed to generate enough electricity for urban areas. Consequently, land for wind farms, especially in coastal areas, can be expensive and difficult to buy from vacationers. Some opponents also dislike the constant, low, humming noise (60 decibels) caused by wind turbines. Table 17.1 shows where wind turbines fall when compared to other sound producing objects and activities.

Hydroelectric Power

Have you ever tried to cross a rushing creek? Even if the water is only a few inches deep, the force of quickly moving water can knock you over. Flowing water creates energy, which can be turned into electricity. This is called *hydroelectric power* or *hydropower*. The two major ways water flow is used to make electricity are (1) huge amounts of water spinning giant turbines, and (2) tidal diversion where water is directed both up and down pipes linked to turbines (i.e., restricting water flow makes the water flow faster, spinning turbines and generating electricity).

Small hydroelectric power systems can provide enough electricity for a home, farm, or ranch. So for those people lucky enough to live near a river, this is a good way to make electricity. Hydroelectric power doesn't pollute the atmosphere like the burning of fossil fuels coal or natural gas. Moving water is powerful and since hydroelectric plants are fueled by water, it's a clean, generally available fuel source.

Types of Hydroelectric Plants

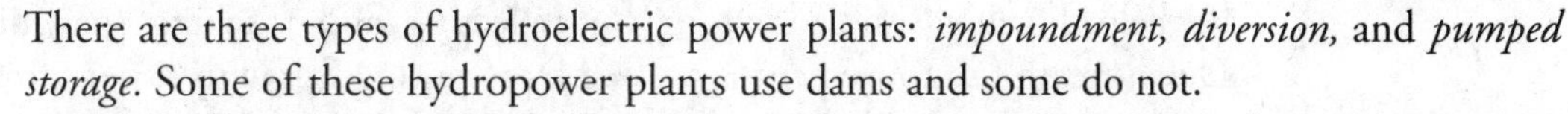

There are three types of hydroelectric power plants: *impoundment, diversion,* and *pumped storage*. Some of these hydropower plants use dams and some do not.

Hydroelectric power is commonly generated at a power plant dam built on a river, where water is stored in a reservoir. When released from the reservoir, water passes through the dam and spins turbines creating electricity. The water is controlled to provide more or less electricity, or maintain the reservoir level. Figure 17.2 illustrates an impoundment (dam) hydroelectric power plant. Hoover Dam near Las Vegas, Nevada, is an example of an impoundment dam.

A *hydroelectric impoundment plant* holds water in a reservoir and then uses the stored potential energy to drive a turbine and produce electricity when the water is released.

Hydroelectric power doesn't always need a big dam. Some hydroelectric power plants use small canals to channel river water through turbines. This type of hydroelectric power

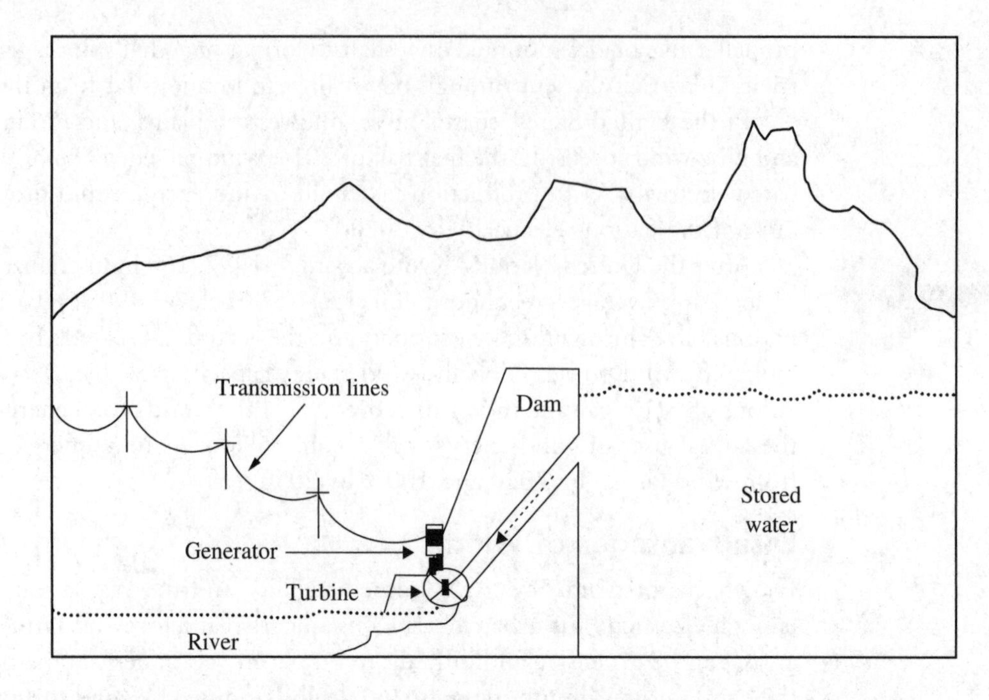

Figure 17.2 A hydroelectric power plant is composed of several parts and processes.

plant is called a *diversion,* or *run-of-river,* plant. It channels a part of a river's flow through a canal or sluice and, depending on the local geography, may not need a dam. Alaska's Tazimina plant is an example of a diversion hydropower plant. No dam is needed.

Some dams, built for irrigation, add a hydroelectric power plant later. In the United States, there are around 80,000 dams. However, only 2,400 produce power. Most dams are for recreation, stock or farm ponds, flood control, water supply, and irrigation. Hydroelectric dams are categorized based on their output of energy. Table 17.2 lists main power divisions.

Hydroelectric Disadvantages

Hydroelectric power plants affect water quality and flow and can cause low dissolved oxygen levels in the water. Drought also affects hydroelectric operations, since plants can't make electricity with low or no water levels. Maintaining normal downstream flow is also important for the survival of marine species and habitats.

Fish are impacted if they aren't able to migrate upstream past impoundment dams to spawning grounds, or if they can't migrate downstream to the ocean. Fish movement can be helped using fish ladders or by collecting and trucking spawning fish upstream. Downstream fish movement is assisted by rerouting fish with screens from turbine intakes, racks, or underwater lights and sounds, and by keeping a very low flow near the turbine.

Table 17.2 There are 3 main classifications of hydroelectric power plants.

HYDROELECTRIC POWER PLANTS	POWER GENERATED
Large	>30 MW
Small hydroelectric power	0.1–30 MW
Micro hydroelectric power	100 kW–0.1 MW

Geothermal Energy

Like solar, the Earth's heat is a gift. The word *geothermal* comes from the Greek words *geo* (earth) and *therme* (heat), and describes the planet's core heat processes from plate tectonics, mountain building, volcanic eruptions, and earthquakes.

The Earth's core temperature is between 4,000°C and 7,000°C. Some internal heat moves upward through tectonic activity and erupts through volcanoes. However, most heat pools into huge underground areas of hot rock, sometimes as big as a mountain range.

The Pacific Ocean is high in geothermal potential. The Atlantic mid-ocean and continental rift zones also have high heat energy (e.g., Iceland and Kenya). Specific hot spots like the Hawaiian Islands and Yellowstone National Park in the United States are also good sources.

Water's Part

Deep subterranean faults and cracks allow rainwater and snow to seep underground. In high-temperature areas, this water is heated and circulates back to the surface. When rising hot water hits solid rock, it's trapped and fills up the holes and cracks of the surrounding rock, forming a geothermal reservoir. Much hotter than surface hot springs, geothermal reservoirs reach temperatures of over 370°C and are huge energy sources.

To capture geothermal energy, holes are drilled into hot spots and the superheated groundwater is pumped to the surface. Scientists and engineers use geological, electrical, magnetic, geochemical, and seismic tests to help find these reservoirs. They are tapped by drilling exploratory wells, sometimes over two miles deep.

Geothermal Power Generation

Many countries operate geothermal power plants producing nearly 10,000 MW of electricity. For example, Kenya on the East African Rift System now generates over eight times more geothermal electricity (202 MW) than does the world's leading energy consumer, China (24 MW). Today, geothermal power provides about 10% of the U.S. energy supply in the western United States.

Electricity generated in the United States from geothermal resources is more than twice that of solar and wind energy combined. However, it's important to know that geothermal energy is not easily transported and loses up to 90% of its heat energy if not used near its source.

A *geothermal district heating* system supplies heat by pumping geothermal water (60°C or hotter) from one or more wells drilled into a geothermal reservoir. Hot water is sent through a heat exchanger, which transfers the heat to water pumped into buildings through separate pipes. After going through the heat exchanger circuit, the used water is directed back into the reservoir to be reheated.

In the western United States, there are over 275 communities close enough to geothermal reservoirs to take advantage of district heating, with 18 district heating systems in use.

Reykjavik, Iceland, has the world's largest geothermal district heating system. Nearly all the homes and buildings use geothermal heat. Before switching to geothermal heat, Reykjavik was heavily polluted from fossil fuel burning. Now, it is one of the cleanest cities in the world.

Disadvantages of Geothermal

Geothermal energy has few polluting problems itself, but there are processing drawbacks. For example, steam can sometimes bring up heavy metals, sulfur, minerals, salts, radon, and toxic gases. If vented above ground in an open-loop system, geothermal energy production can pollute. Scrubbers filter toxic components, but produce hazardous sludge and can potentially contaminate soil and groundwater. In a closed-loop (recycled) system, pollutants don't come above ground.

Another geothermal drawback involves location, since some geothermal areas, like Yellowstone National Park, are located in pristine, environmentally sensitive areas. Careless construction of geothermal plants could greatly impact local ecology.

Ocean Tidal Energy

For coastal inhabitants and industries, the ocean offers another renewable energy resource. Incoming and outgoing tides can be tapped with structures called tidal barrages. Placed across tidal basins, water movement is used to turn turbines and generate electricity. Although initial construction costs may be high, operating costs are low.

Hydrogen Cells

Hydrogen is considered to be the safest and cleanest of energy sources. Hydrogen is released during *electrolysis* when hydrogen atoms cleave from a water molecule. Then, the hydrogen atoms are stored and used to create electricity through a reversal of the electrolytic process.

Unlike other energy sources, hydrogen fuel cells give off steam as their only waste. However, hydrogen production costs are currently high. For hydrogen to become a viable fuel source, a much cheaper production method will be needed.

Renewable Energy Certificates

Renewable energy certificates (*green certificates* or *green tags*) describe environmental characteristics of power from renewable energy projects and are sold separately from general electricity. Consumers can buy green certificates whether or not they have access to green power through their local utility. They can also buy green certificates without having to switch electricity suppliers. Today, over 30 organizations sell wholesale or retail green energy certificates.

Environmental Resources Trust Inc. (ERT) is a nonprofit organization based in Washington, D.C., that uses market forces to protect and improve the global environment. Established in 1996, ERT uses energy markets to meet the challenges of climate change, secure clean and reliable power, and encourage sustainable land use.

ERT initiated three focused programs to carry out its mission. The GHG Registry validates industrial greenhouse gas emission profiles by creating a market that can facilitate emission decreases. The EcoPower Program verifies and promotes blocks of clean power from new renewable energy sources. The EcoLands Program works out plans to encourage and assist landowners in land use. The bulk of ERT's Clean Power Program involves

1. Verifying specific energy blocks as green and giving them the EcoPower label
2. Creating a *Power Scorecard Rating System* that categorizes the environmental attributes of various clean power blocks
3. Advertising EcoPower blocks and negotiating their sale from generators to consumers, like municipalities
4. Developing an EcoPower ticket program guaranteeing consumers a definite claim to purchased power
5. Auditing EcoPower energy blocks and preparing verification reports

ERT's Power Scorecard Rating System, verification, and marketing have raised public awareness of the clean power market. EcoPower tickets make it possible for clean energy generated in one area to be available for sale elsewhere. Today, legitimate energy providers who advertise clean energy to consumers have each kilowatt-hour verified as "new, certified, zero-emissions renewable power" by ERT.

› Review Questions

Multiple-Choice Questions

1. The Staebler-Wronski effect describes

   (A) turbine wind shear factor
   (B) greenhouse gas warming
   (C) geothermal heat exchange
   (D) solar cell degradation
   (E) hydroelectric heat loss

2. Geothermal reservoirs can reach temperatures of

   (A) 370°C
   (B) 560°C
   (C) 740°C
   (D) 820°C
   (E) 1,060°C

3. Impoundment, diversion, and run-of-river are all types of

   (A) solar cells
   (B) wind turbines
   (C) hydroelectric power plants
   (D) canoes
   (E) nuclear power plants

4. Ethanol is used in the United States as a fuel additive to

   (A) increase fuel safety
   (B) keep the carburetor new
   (C) supplement grain production
   (D) reduce incomplete combustion emissions
   (E) make gasoline last longer

5. Flowing water that creates energy and is turned into electricity is called

   (A) nuclear power
   (B) hydroelectric power
   (C) solar power
   (D) thermal energy
   (E) chemical energy

6. Turbines catch the wind with two to three propeller-like blades mounted on a shaft to form a

   (A) rotor
   (B) heat exchanger
   (C) reservoir
   (D) magnetic field
   (E) storage battery

7. What percent of the world's population uses wood or charcoal as a main energy source?

   (A) 15%
   (B) 20%
   (C) 40%
   (D) 50%
   (E) 90%

8. When deep underground heat is transferred by thermal conduction through water to the surface, it is called

   (A) solar energy
   (B) nuclear energy
   (C) wind energy
   (D) geothermal energy
   (E) cosmic energy

9. Propeller-like turbine blades are used to generate electricity from

   (A) sunlight
   (B) rain
   (C) wind
   (D) snow
   (E) biomass

10. What percent of U.S. energy needs are met by geothermal power?

    (A) 10%
    (B) 20%
    (C) 25%
    (D) 35%
    (E) 50%

11. A Power Scorecard Rating System categorizes

    (A) energy availability
    (B) geothermal electricity requirements
    (C) turbine rotor speed
    (D) environmental aspects of clean power blocks
    (E) public acceptance of tidal energy

12. Tidal energy taps

    (A) stream flow
    (B) incoming and outgoing tides
    (C) salinity levels
    (D) rising geothermal magma
    (E) melting icebergs

13. Industrial greenhouse gas emission profiles are products of which nonprofit organization?

(A) International Oil Tanker Owner's Pollution Federation
(B) GHG Registry
(C) Superfund
(D) Occupational Safety and Health Agency
(E) EcoPower

14. Diffusing phosphorus onto boron-coated silicon wafers produces

(A) acetate
(B) hydroelectric conduits
(C) semiconductor connections
(D) radioactive markers
(E) an electricity grid

15. All the following are good sites for wind turbines except

(A) forests
(B) coastal areas
(C) plains
(D) mountain passes
(E) open ocean

16. The problem with ethanol is its ability to

(A) cause soil erosion
(B) pollute drinking water
(C) evaporate into the atmosphere
(D) increase the polluting effects of other compounds
(E) reduce organic compounds

17. What element is released during electrolysis?

(A) Sulfur
(B) Nitrogen
(C) Helium
(D) Radon
(E) Hydrogen

18. What safe fuel offers superior performance and is the fuel of choice for race cars?

(A) Hydrogen
(B) Methanol
(C) Gasohol
(D) Ethanol
(E) Propane

19. The Earth gets more energy from the sun in an hour than

(A) New York City uses in a month
(B) Asia uses in a week
(C) the planet uses in a year
(D) the Pacific Ocean can absorb
(E) photovoltaic cells can tolerate

❯ Answers and Explanations

1. **D**—The Staebler-Wronski effect describes solar cell degradation (i.e., the thinnest layers increase the electric field strength across the material, but reduce light absorption and cell efficiency).
2. **A**—Geothermal reservoirs can reach temperatures of 370°C.
3. **C**—Impoundment and diversion or run-of-river are types of hydroelectric dams.
4. **D**—Ethanol is used in the United States as a fuel additive to reduce incomplete combustion, burn cleanly, and reduce greenhouse gases.
5. **B**—Flowing water (*hydro*) creates energy and is turned into electricity.
6. **A**—Turbines catch the wind with two or three propeller-like blades mounted on a shaft to form a rotor attached to a power generator.
7. **C**—Forty percent of the world's population uses wood or charcoal as their main energy source.
8. **D**—Deep underground heat, transferred by thermal conduction through water to the surface, is called geothermal (Earth's heat) energy.
9. **C**—Airplane-like propeller blades are used to generate electricity from wind turbines.
10. **A**—Because of location, 10% of U.S. energy comes from geothermal power.
11. **D**—A Power Scorecard Rating System categorizes environmental aspects of verified clean power blocks from new renewable energy sources.
12. **B**—Tidal energy taps incoming and outgoing water flow during high and low tides.
13. **B**—Industrial greenhouse gas emission profiles are products of the GHG Registry.
14. **C**—Diffusing phosphorus onto boron-coated silicon wafers produces semiconductor connections.
15. **A**—Coasts, plains, mountain passes, and open oceans are good for wind turbines, but forests tend to block or disperse wind.
16. **D**—The problem with ethanol is its ability to increase benzene and toluene's time in the environment before they break down.
17. **E**—Hydrogen is released as a result of hydrolysis.
18. **B**—Because of clean burning, methanol provides race cars with superior performance and safety.
19. **C**—The Earth gets more energy from the sun in an hour than the planet uses in a year (i.e., roughly 10,000 times all the commercial energy produced yearly).

Free-Response Questions

1. The European Wind Energy Association plans to provide over 10% of Europe's electricity by 2030. Denmark's wind energy program shows how government support can assist in making green energy sources commercially viable. Its government has set the target of 50% of electricity provided by wind energy by 2020.

   In the United States, 11,329 MW of wind energy were generated in 2006, with additional wind capacity coming on line. With an energy cost of 2.0 cents/kWh, wind is equal to or less expensive than coal, oil, nuclear, and most natural gas–fired generation. The great thing about wind, besides being a clean energy source, is that it's free after initial construction.

   (a) Describe possible disadvantages of wind-generated electricity.
   (b) List the long-term benefits of wind energy.

2. ERT is made up of the National Audubon Society, the National Fish and Wildlife Foundation, Environmental Defense, and the German Marshall Fund. ERT energizes green power markets by supplying important auditing and verification services. These services support consumer confidence by verifying that new green power sources are actually being brought on line.

(a) Describe how the Power Scorecard Rating System works.
(b) How do EcoPower blocks benefit consumers?

Free-Response Answers and Explanations

1.

a. Wind energy disadvantages are mainly socioeconomic. Though the monetary costs of building wind turbines is going down, wind energy is not yet an economically efficient way to produce electricity on a mass scale. While wind is readily available all over the world, it is both unpredictable and not strong enough in all areas as a source of electricity. Creating the necessary number of turbines to produce electricity for urban areas requires extensive land use and creates noise pollution for nearby populations. Wind farms also detract from the natural beauty of their locales, thus devaluing a land's scenic value. This could become an economic burden rather than boon, particularly in coastal areas where tourism provides a much needed source of revenue.

b. Wind energy is a clean, renewable energy resource that does not contribute to the creation of greenhouse gases like CO_2. Using wind energy places much less stress on the surrounding environment than other energy production methods. Wind energy also has long-term social and economic benefits. Because its "fuel" is free and inexhaustible, its use can reduce and therefore help stabilize the demands on other valuable resources such as natural gas and oil. Wind energy can generate electricity for remote areas marginalized by geographic locale, and improve the quality of life for populations all over the world. Finally, because of the low environmental impact of turbines, the land beneath turbines can be used for multiple purposes (e.g., agriculture), allowing maximum land use.

2.

a. Developed by the Environmental Resources Trust, the Power Scorecard Rating System categorizes electricity products according to their environmental impact and commitment to new renewable energy. This ratings system creates a value structure that places a premium on new renewable energy products that minimize environmental impact. It helps ERT harness the power of the economy to not only raise public awareness of new, clean energy resources, but stimulate clean energy business allowing market forces to provide businesses the incentives to use clean energy.

b. One benefit EcoPower blocks give to consumers is the ability to "speak" through economic means. By informing consumers about clean energy products and resources, EcoPower blocks help bridge the gap between existing clean energy producers and lack of public information. EcoPower blocks clearly define and provide assurances about clean energy products. The EcoPower program harnesses market forces to drive positive global environmental changes and gives consumers the ability to promote clean energy through buying power.

❯ Rapid Review

- The Staebler-Wronski effect describes solar cell degradation (i.e., the thinnest layers increase the electric field strength across the material, but reduce light absorption and cell efficiency).
- Geothermal reservoirs can reach temperatures of 370°C.
- Impoundment and diversion or run-of-river are types of hydroelectric dams.
- Ethanol is used in the United States as a fuel additive to reduce incomplete combustion, allow the fuel to burn cleanly, and reduce greenhouse gases.
- Flowing water (*hydro*) creates energy and can be turned into electricity.
- Turbines catch the wind with two to three propeller-like blades mounted on a shaft to form a rotor attached to a power generator.
- Forty percent of the world's population uses wood or charcoal as a main energy source.
- Deep underground heat, transferred by thermal conduction through water to the surface, is called geothermal (Earth's heat) energy.
- Airplane-like propeller blades are used to generate electricity from wind turbines.
- Taking advantage of location, 10% of U.S. energy comes from geothermal power.
- A Power Scorecard Rating System categorizes environmental aspects of verified clean power blocks from new renewable energy sources.
- Tidal energy taps incoming and outgoing water flow during high and low tides.
- Industrial greenhouse gas emission profiles are products of the GHG Registry.
- Diffusing phosphorus onto boron-coated silicon wafers produces semiconductor connections.
- Coasts, plains, mountain passes, and open oceans are good for wind turbines, but forests tend to block or disperse the wind.
- The problem with ethanol is its ability to increase benzene and toluene's time in the environment before they break down.
- Hydrogen is released as a result of hydrolysis.
- Because it burns cleanly, methanol provides race cars with superior performance and safety.
- The Earth gets more energy from the sun in an hour than the planet uses in a year (i.e., roughly 10,000 times all the commercial energy produced yearly).

Pollution Types

IN THIS CHAPTER

Summary: The Earth's atmosphere and water make up a delicately balanced system that functions well until disrupted by outside forces or contaminants. Some natural contaminants are naturally cleared, while others from human activities take much more effort to eliminate.

Keywords

✪ Ozone, troposphere, volatile organic compounds (VOCs), point source, nonpoint source, smog, heat island, dissolved oxygen, dead zones, Superfund, fecal coliform bacteria, total organic carbon (TOC), thermal pollution, sick building syndrome

Atmosphere

Our atmosphere contains oxygen from algae and plants, but the primeval atmosphere was mostly volcanic gases with little oxygen. Today, there are four distinct layers (i.e., troposphere, stratosphere, mesosphere, and thermosphere) divided by temperature, chemical properties, and gaseous mixing. Atmospheric gases include 80% nitrogen (by volume), 20% oxygen, 0.036% carbon dioxide, and trace amounts of other gases. Refer back to Figure 6.1 for these layers.

Virtually all living organisms and human activities occur in the troposphere, which is protected from harmful incoming radiation. Rising and falling temperatures, as well as circulating air masses, keep things lively. When compared to the other layers, however, the troposphere is thin.

Air Pollution

Urban *smog* is a major air pollutant regulated by the EPA, but it is not emitted directly from specific sources. It's formed in the atmosphere from nitrogen oxides and *volatile organic compounds (VOCs)*. Sources of VOCs include: (1) combustion products from motor vehicles and machinery that burns fossil fuels, (2) gasoline vapors from cars and fueling stations, (3) refineries and petroleum storage tanks, (4) chemical solvent vapors from dry-cleaning processes, (5) solid waste facilities, and (6) metal-surface paints. Internal combustion engine fumes contain many VOCs that, when released into the atmosphere, interact with other gases and sunlight to create the ozone part of smog. The EPA has targeted VOC reduction as an important control mechanism for reducing high-ozone-containing smog in cities.

Since reactions forming ozone are affected by sunlight, high ozone levels usually occur in the summer months when the air is hot and slow moving. In the summer, more people are also traveling in addition to their daily commute, so vehicle emissions rise.

Topography and Heat

When a stable layer of warm air rolls over the top of cooler air, a *temperature inversion* occurs. Instead of it getting colder higher in the atmosphere, it gets warmer. This is particularly true in cities rimmed by mountains (e.g., Los Angeles). At night the land cools, but because pollutants hold the day's heat, cool air slips below and causes an inversion. During the day, air currents mix in more pollutants and by late afternoon a brown haze (i.e., smog) makes eyes water and sinuses burn. At this point, the air is a health hazard.

Smog irritation of the mucous membranes of the nose, throat, and lungs depends on ozone levels, as well as the frequency and duration of exposure. In fact, when urban ozone concentrations are high, illness and hospital admissions go up.

Huge cities with tall buildings and miles of concrete and glass add to this problem. With little plant life to absorb the heat, cities produce high heat gradients during the day and release heat at night. They are often 3–5°C higher than surrounding areas.

These city *heat islands* collect pollutants so that areas downwind have much less visibility and greater rainfall due to condensation characteristics than their neighbors with cleaner air. City aerosols and dust also trigger lightning strikes. In fact, Houston, Texas, which has many oil refineries and chemical plants, has a relatively high number of lightning strikes compared to most other areas in the United States.

Indoor Air

Industry has developed amazing products (e.g., complex carpet fibers, composite wood, linoleum, plastics, paints, and solvents) in the past 60 years. These products make our lives easier, but often contain toxic compounds (e.g., formaldehyde, a known *carcinogen*).

When homes are well ventilated, it isn't a problem, but energy-efficient homes are airtight. Toxic pollutants are trapped indoors where we spend the majority of our time. In the workplace, bad air mixed with mold spores has led to *sick building syndrome.* People suffer headaches, allergies, fatigue, nausea, and respiratory problems leading to greater medical costs, days off, and low productivity.

When the Clean Air Act named sulfur dioxide, carbon monoxide, nitrogen oxides, photochemical oxidants, particulates, volatile hydrocarbons, and lead as health threats in 1970, it opened the door for stricter regulation of polluting industries. For example, the *Energy Policy and Conservation Act* of 1975 allowed the U.S. Department of Transportation to set the Corporate Average Fuel Economy (CAFE) for motor vehicles. To reduce fuel consumption and emissions, the standard requires vehicles to have an average fuel efficiency of

27.5 mi/gal, with larger vehicles (e.g., SUVs and minivans) at 22.5 mi/gal. Tier 2 standards of 2007 require light trucks to meet the stricter passenger car standard.

CAFE Tier 2 standards also reduced nitrous oxide emissions to 0.07 grams per mile (i.e., 90% reduction) for cars, and targeted gasoline sulfur emissions to drop from 300 to 30 parts per million (ppm).

In 2005, President George W. Bush announced the *Clean Air Interstate Rule* (CAIR), which set sulfur dioxide, nitrogen oxides, and particulate emissions limits. By 2018, this will decrease these chemicals by 70%.

In addition to removing particulates and limiting industrial production, citizens can also help by conserving energy, using nonpolluting four-cycle gasoline engines, planting trees, writing Congressional representatives, installing high-efficiency fireplace inserts, using latex paints, and reducing dry cleaning.

Water

Pure water (H_2O), completely free from any dissolved substances, is found only in the laboratory. Natural water contains dissolved gases and salts. For example, water must contain enough dissolved oxygen for fish to survive or they die. Drinking water, without dissolved oxygen or dissolved mineral salts, tastes bad. Salts give water its taste.

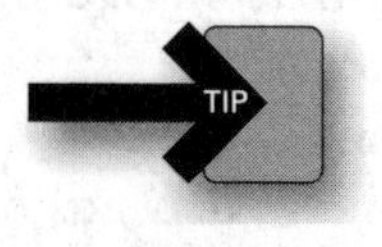

A *total organic carbon* (*TOC*) level is measured by hydrologists when checking the health of freshwater. As we have seen, organic matter plays a big role in aquatic systems. It affects biogeochemical processes, nutrient cycling, biological availability, chemical transport, and interactions. It also directly affects municipal choices for wastewater and drinking water treatments. Organic content is commonly measured as total organic carbon and dissolved organic carbon, which are essential components of the carbon cycle.

Pollution

Water pollution is caused by the sudden or ongoing, accidental or deliberate, discharge of a polluting material. Increasing human populations put pressure on the oceans and marine environment. More and more people on the planet lead to more

- Sewage produced
- Fertilizers, herbicides, and pesticides used for crops, lawns, golf courses, and parks
- Fossil fuels extracted and burned
- Oil leaked and spilled
- Land deforested and developed
- Various by-products of manufacturing and shipping generated

Cultural, political, and economic forces affect the types, amounts, and management of waste produced. Increasing population is just one contributor to increasing pollution. As with everything in the environment, the causes and effects are complex.

Water pollution comes from the loss of any real or potential water uses caused by a change in the water's composition due to human activity.

Water is used for everything from drinking and household needs to watering livestock and the irrigation of crops. Fisheries, industry, food production, bathing, recreation, and other services all use water to a large extent. When water becomes unusable for any of these purposes, it is polluted to a greater or lesser degree depending on the extent of the damage caused.

Groundwater has been contaminated by leaking underground storage tanks, fertilizers and pesticides, unregulated hazardous waste sites, septic tanks, drainage wells, and other sources, threatening the drinking water of 50% of the U.S. population.

The three major sources of water pollution are *municipal, industrial,* and *agricultural.* Municipal water pollution comes from residential and commercial wastewater. In the past, municipal wastewater was treated by reducing suspended solids, oxygen-demanding materials, dissolved inorganic compounds, and harmful bacteria. Today, the focus is on improving solid residue disposal from municipal treatment processes.

Agricultural areas in the United States have also developed water pollution problems. For example, in Iowa where chemical fertilizers are used across 60% of the state, private and public drinking water wells have exceeded safety standards for nitrates. Towns in Nebraska have also shown high nitrate levels and require monthly well testing.

Runoff

Pollution of marine ecosystems includes runoff from land, rivers, and streams; direct sewage discharge; air pollution; and discharge from manufacturing, oil operations, shipping, and mining.

Although coastal cities have the greatest impact on ocean ecosystems, pollution from runoff is not limited to coastal regions. Runoff from over 90% of the Earth's land surface (inland and coastal) eventually drains into the sea, carrying sewage, fertilizers, and toxic chemicals. Similarly, air pollution from inland as well as coastal cities, including byproducts of fossil fuel consumption, polychlorinated biphenyls (PCBs), metals, pesticides, and dioxins, eventually finds its way into the oceans after rain or snow.

Oil Spills

Increased oil demand has increased offshore oil drilling operations and oil transport. These activities have resulted in many oil spills. The number of oil spills worldwide of between 7 and greater than 700 tons has varied in the past 30 years, with some years better than others. Table 18.1 lists the top 15 oil spills (excluding acts of war) recorded by the *International Oil Tanker Owners Pollution Federation* (*IOTPF*). The IOTPF measures all oil lost to the environment, including oil burned and released into the atmosphere or still in sunken ships. Relatively speaking, the well-known *Valdez* spill of 1989 off the coast of Alaska, and thought by many to be a particularly bad spill, was not as extensive as many others (it rates 35th in largest spills), but had a huge impact on the delicate and pristine arctic environment.

Spills account for only 10% of marine oil pollution. About 50% of oil pollution in marine waters comes from ongoing low-level sources such as marine terminal leaks, dumping of offshore oil drilling mud, land runoff, and atmospheric pollution from incompletely burned fuels.

Dead Zones

Cumulative pollution effects on ocean ecosystems are very serious. For example, in the Gulf of Mexico, scientists have identified *dead zones* in once highly productive waters. These zones have been traced to excessive nutrients from farms, lawns, and inadequately treated sewage. This stimulates rapid plankton growth that in turn leads to oxygen depletion in the water.

Blooms of toxic *phytoplanktons* and red tides have increased in frequency over the last two decades and may be linked to coastal pollution. For example, storm water runoff contains suspended particulates, nutrients, heavy metals, and toxin. The effects of storm water runoff often cause *dinoflagellate* (red tide) blooms following storms. These tides cause high numbers of fish and marine mammal deaths and can be a serious threat to human health.

Table 18.1 Oil spill impacts are as related to spill location as total volume lost.

SHIP	YEAR	LOCATION	VOLUME (tonnes)
MV *Miss Susan*/MV *Summer Wind* collision	2014	Houston Ship Channel, Houston, TX, United States	546
North Dakota train collision	2013	Casselton, North Dakota, United States	1,300
Lac-Mégantic derailment	2013	Québec, Canada	4,830
Guarapiche	2012	Guarapiche River, Maturin, Venezuela	41,000
Rena (cargo ship)	2011	Tauranga, New Zealand	350
Deepwater Horizon/BP (well blowout and spill)	2010	Gulf of Mexico	627,000
Montara oil field	2009	Timor Sea, Australia	30,000
Hurricane Ike (oil platforms, tanks, pipelines)	2008	Gulf of Mexico	8,800
Hebel Spirit	2007	South Korea	8,812
Hurricane Katrina (oil platforms, tanks, pipelines)	2007	Gulf of Mexico	22,029
Solar 1	2006	Guimaras Island, Phillipines	1,540
Prestige (sank with oil inside)	2004	Galicia, Spain	62,941
MV *Selendang*	2004	Aleutian Islands, Alaska	1,560
Treasure	2000	Cape Town, South Africa	1,400
Erika	1999	Coast of Brittany, France	25,000
Fergana Valley	1992	Uzbekistan	285,000
Exxon *Valdez*	1989	Prince William Sound, Alaska, United States	104,000
Odyssey	1988	St. John's, Newfoundland	132,000

pH

Measurement of the *pH* of wastewater is an important factor used in decisions related to its treatment and eventual release into natural water ecosystems.

The measurement of the number of hydrogen ions in water, on a scale from 0 to 14, is called the water's *pH.*

A solution with a pH value of 7 is neutral, while a solution with a pH value less than 7 is acidic and a solution with a pH value greater than 7 is basic. Natural waters usually have a pH between 6 and 9. The scale is negatively logarithmic, so each whole number (reading downward) is 10 times the preceding one (for example, pH 5.5 is 100 times as acidic as pH 7.5).

$$pH = -\log [H+] = \text{hydrogen ion concentration}$$

The pH of natural waters becomes acidic or basic as a result of human activities such as acid mine drainage, emissions from coal burning power plants, and heavy automobile traffic.

Dissolved Oxygen

Oxygen enters water by direct atmospheric absorption or by aquatic plant and algal photosynthesis. Oxygen is removed from water by respiration and the decomposition of organic material.

Dissolved oxygen is the amount of oxygen measured in a stream, river, or lake.

Dissolved oxygen is also an important marker of a river or lake's ability to support aquatic life. Fish need oxygen to survive and absorb dissolved oxygen through their gills. Dissolved oxygen present in even the cleanest water is extremely small and depends on several factors, including temperature (e.g., the colder the water, the more oxygen that can be dissolved), water flow volume and velocity, and the number of organisms using oxygen for respiration.

Oxygen solubility in water at a temperature of 20°C is 9.2 milligrams oxygen per liter of water (about 9 parts per million).

Dissolved oxygen in water is expressed as a concentration in milligrams per liter (mg/L) of water. Metropolitan activities affecting dissolved oxygen levels include removal of native vegetation, runoff, and sewage discharge. Many of these pollutants are nontoxic, so how do they cause pollution problems?

The answer comes back to the oxygen levels. A rapidly flowing stream reaches 100% saturation of around 9 ppm, which allows healthy growth of natural flora and fauna (e.g., animals and bacteria). The bacteria are mostly aerobic (they require oxygen), and their numbers are controlled by the availability of food (digestible organic matter). Bacterial growth is greatly stimulated when there is a big discharge of organic materials (e.g., sewage, milk, or agricultural waste). Bacteria populations grow rapidly, consuming and depleting water oxygen levels.

The amount of oxygen depletion over time depends on the speed with which the stream takes up oxygen from the atmosphere (*reaeration capacity*). Fast-flowing streams reoxygenate quickly, while deep, slow-flowing rivers take up oxygen slowly. Oxygen loss may be counteracted by plant photosynthesis, which produces oxygen during daylight. However, plant processes can't keep up in heavily polluted areas and oxygen levels drop quickly. In anaerobic conditions (i.e., complete lack of free oxygen), fish die.

Water Treatment

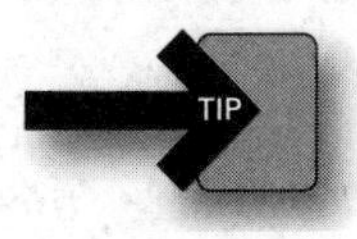

Whenever water is used for humans, it must be treated from two different angles. First, any surface water from rivers that is used in cities is treated for drinking; usually by chlorination. After water is used for drinking, washing, lawns, toilets, and so forth, it has to be treated at a wastewater treatment plant before it can be released back into the environment. Most municipal water purification systems use several steps to treat water, from physical removal of surface impurities to chemical treatment. Figure 18.1 illustrates the path water takes from initial water treatment (*chlorination*) to urban use, and then to *wastewater treatment* before its release back into the environment.

Before raw water is treated, it passes through large screens used to remove sticks, leaves, and other large objects like plastic bottles. Sand and grit settle out or fall to the bottom of a tank during this stage. During *coagulation,* a chemical such as aluminum sulfate is added to the raw water, forming sticky blobs that snag small particles of bacteria, silt, and other impurities. *Flocculation* removes impurities by skimming the top of the tank. Water is then pumped slowly through a long basin called a *settling basin.* This is done to remove much of the remaining solid material, which collects at the bottom of the basin during *sedimentation* or *clarification.*

Next, microorganisms like viruses, bacteria, and protozoa, as well as any remaining small particles, are removed. This is done by water *filtration* through layers of sand, coal, and other granular materials. After the water is filtered, it is treated with chemical disinfectants to kill any organisms not collected during filtration.

Chlorine, often the only chemical treatment method used on surface water, is not without problems. When chlorine mixes with organic material, it creates potentially dangerous *trihalomethane* (THM). Big treatment plants remove THM to safe levels, but small towns often don't.

Ozone oxidation is another good disinfectant method, but unlike chlorine, ozone doesn't stay in water after it leaves the treatment plant. So bacteria lurking in municipal or residential water pipes aren't killed.

Ultraviolet light has also been used to treat wastewater by killing microorganisms, but like oxidation, it is a one-time treatment at the plant. There is no continuing protection like that provided by chlorine, but people who oppose water chlorination prefer it.

Turbidity is a measure of water's cloudiness; the cloudier the water, the higher the turbidity. Water turbidity, caused by suspended matter such as clay, silt, and organic matter, can also result from microscopic organisms blocking light through water. Though not a major health concern, turbidity blocks disinfection and augments microbial growth. High turbidity can also be caused by soil erosion, urban runoff, and high flow rates.

Wastewater Treatment

Wastewater treatment starts with screening for large particles, followed by an aerobic system with activated sludge to remove organics. Next, sedimentation and removal of organic biomass takes place, often recycling more than once. Biomass sludge is removed from wastewater before the filtration step. Wastewater filtration, followed by disinfection with chlorine and its removal, takes place before clean water is finally discharged into the environment.

Contaminants

Pollution is bad news. It poisons drinking water and land and marine animals (through bioaccumulation), upsets aquatic ecosystems, and causes deforestation through acid rain.

In general, four main water contaminants exist: *organic, inorganic, radioactive,* and *acid-base.* Released into the environment in different ways, most pollutants enter the hydrologic cycle as direct (*point source*) and indirect (*nonpoint source*) contamination.

Point sources (e.g., water from factories, refineries, and waste treatment plants) are released directly into urban water supplies. In the United States and elsewhere, these releases are regulated, but some pollutants are still found in these waters.

Nonpoint sources include contaminants entering the water supply from soils and groundwater systems runoff and rainfall. Soils and groundwater contain fertilizer and pesticide residues, as well as industrial wastes. Atmospheric contaminants also come from gaseous emissions from automobiles, factories, and even restaurants.

In 1987, the U.S. Environmental Protection Agency (EPA) issued the *Clean Water Act,* Section 319, calling for federal cooperation, leadership, and funding to help state and local governments tackle nonpoint-source pollution.

Chemicals

Many pollution sources contain high nutrient levels, which affect the photosynthetic cycle of water plants and organisms. This hurts fish and shellfish living in the water.

Nitrogen

Nitrogen, needed by all organisms to grow and reproduce is very common, and found as *nitrate* (NO_3), *nitrite* (NO_2), *ammonia* (NH_3), and *nitrogen gas* (N_2). Organic nitrogen in the cells of living things makes up proteins, peptides, and amino acids. Nitrogen enters waterways from lawn fertilizer runoff, leaking septic tanks, animal wastes, industrial wastewaters, sanitary landfills, and car exhausts.

Phosphorus

Phosphorus is a nutrient needed by all organisms for basic biological processes. It is found in mineral deposits, soils, and organic material, and in very low concentrations in fresh water. However, phosphorus used widely in detergents, fertilizer, and other products is often found in higher concentrations (e.g., phosphate) in the surface waters of populated areas.

Hazardous Waste

When hazardous waste sites are discovered years after they have been abandoned, it is often individuals or communities who have experienced most of the ill effects (e.g., polluted water, increased birth defects). Sometimes contaminants were dumped illegally or land was purchased for waste disposal and then years later, homes and schools were built on the toxic sites. This is what happened in 1978 at Love Canal, New York. It wasn't until residents noticed dying trees, barrels leaking toxic liquids, and smelly sludge in their basements that the EPA was called.

The land's past use was investigated and families had to be moved away. The entire Love Canal area was closed and cleaned up, costing millions of dollars. This resulted in federal legislation in 1980 called the *Comprehensive Environmental Response Compensation and Liability Act* (CERCLA) or, as it is commonly called, the *Superfund.*

The Superfund established a federal tax on chemical and oil industries and created rules on how abandoned hazardous sites are handled. The Superfund also made it possible to go after polluting offenders and create a trust fund for cleanup when no offender could be found.

Organic Matter

Pollution occurs when silt and other suspended solids (e.g., soil) run off newly plowed fields, construction and logging sites, residential areas, and river banks after a rain or snow melt. In the presence of excess phosphates, lakes and slow-moving rivers go through *eutrophication,* gradually filling them in with sediment and organic matter. When extra sediments enter a lake, fish respiration is impacted, plant growth and water depth are limited, and aquatic organisms asphyxiate. Early phosphate removal helps prevent eutrophication.

Organic pollution enters waterways as sewage, leaves and grass clippings, or as runoff from livestock feedlots and pastures. When bacteria break down this organic material, measured as *biochemical oxygen demand* (*BOD*), they use oxygen dissolved in the water. Most fish can't live when dissolved oxygen levels drop below 2 to 5 parts per million. So, when this happens, fish and other water organisms die in huge numbers and the food web is hit hard. Pollution of rivers and streams is one of the most critical environmental problems of the past 100 years causing a domino effect of destruction.

Measuring Toxicity

Subtle differences in the human genome may explain why what affects some individuals greatly has little or no affect on others. For this reason when measuring toxicity and setting safety limits, individual and species variations must be taken into account. Some species are more susceptible to specific chemicals than others. For example, of the 226 known cancer-causing chemicals in rats and mice, 95 cause cancer in one species, but not the other.

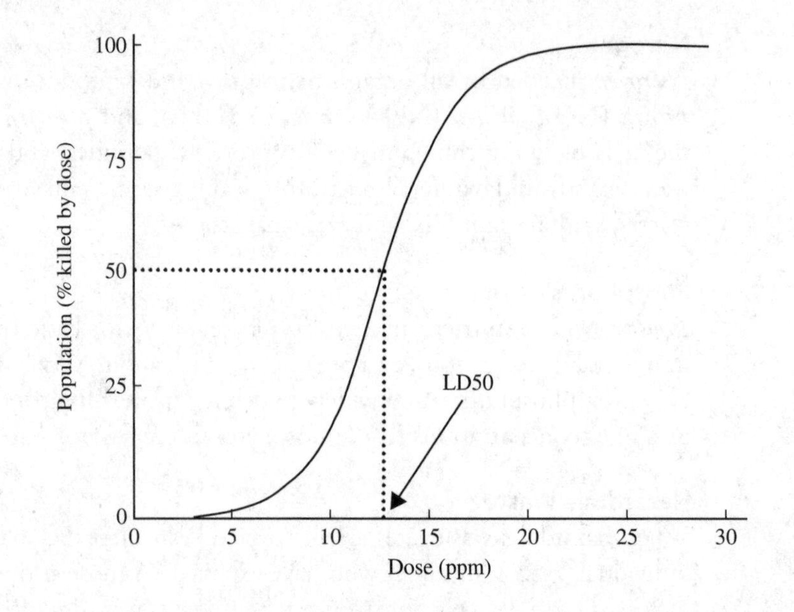

Figure 18.1 Scientists use dose response curves to set acceptable limits for pollutants/carcinogens.

When setting acceptable limits for different pollutants, scientists use *dose response curves.* A chemical's toxicity dose is equal to that amount at which 50% of a test population is sensitive. A lethal dose is written as *LD50.* Figure 18.1 shows a hypothetical chemical's LD50 and dose response curve.

Pathogens

Fecal coliform bacteria, present in the feces and intestinal tracts of humans and warm-blooded animals, enter rivers and lakes from human and animal waste. When fecal coliform bacteria are present, it is an indication of pathogenic microorganisms.

Pathogens or disease-causing microorganisms cause everything from typhoid fever and dysentery to respiratory and skin diseases. Microscopic pathogens (e.g., bacteria, viruses, and protozoa) enter waterways through untreated urban sewage, storm drains, pet waste, septic tanks, farm runoff, and bilge water, causing sickness and/or death.

Protecting human health is the key concern in water treatment. Removal of potential pathogens, turbidity, hazardous chemicals, and nitrates are all important factors in keeping water safe. Poor and war-torn countries around the world have no clean drinking water. The vast majority of disease in these countries is directly related to polluted water supplies.

Acid Rain

Rain is naturally acidic (pH of 5.6 to 5.7) because water reacts with carbon dioxide in the atmosphere to form carbonic acid. Acid rain is formed when elevated levels of atmospheric chemicals (i.e., nitrogen, sulfur, and carbon) react with water and turn to the Earth in acidic raindrops.

When acid rain falls on limestone statues, monuments, and gravestones, it dissolves, discolors, and/or disfigures the surfaces by reacting with the rock. This is known as *dissolution.* Statues and buildings, hundreds to thousands of years old, suffer this kind of weathering.

As early as the 17th and 18th centuries, acid rain affected plants and people. Angus Smith published a book called *Acid Rain* in 1872. However, it wasn't until fishermen saw fish numbers and diversity declining throughout North American lakes and Europe that it became globally recognized. The eastern North American coast has precipitation pH levels near 2.3 or about 1,000 times more acidic than pure water.

Acid rain refers to all types of precipitation (rain, snow, sleet, hail, fog) that are acidic (pH lower than the 5.6 average of rainwater) in nature.

Sulfur and nitrogen oxides in acid rain are released from industrial smokestacks, vehicle exhausts, and wood burning. In the atmosphere, oxides mix with moisture becoming sulfuric and nitric acid and fall to the ground as rain and snow.

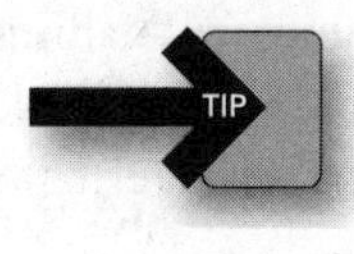

Acid rain has been measured in the United States, Germany, the Czech Republic, the Netherlands, Switzerland, and Australia. It is also becoming serious in Japan, China, and Southeast Asia. Acid rain affects lakes, streams, rivers, bays, ponds, and other bodies of water by increasing acidity until aquatic creatures can no longer live. Aquatic plants grow best between pH 7.0 and 9.0. As acidity increases, submerged plants die and waterfowl lose a basic food source. At pH 6, freshwater shrimp cannot survive. At pH 5.5, bacterial decomposers die, organic debris stops getting broken down and builds up on the bottom, and plankton die off. Below 4.5 pH, all fish die.

Acid rain also harms surface vegetation. Forests in western Europe are thought to be dying from acid rain. Scientists think acid rain damages leaves' protective coating, allowing acids to penetrate. This disrupts water evaporation and gas exchange to the point that the plant can no longer breathe, convert nutrients, or take up water.

A big acid rain effect on forests is nutrient leaching from the soil and toxic metal concentration. Nutrients leach out when acid rain adds hydrogen ions to the soil and reacts with local minerals (e.g., calcium, magnesium, and potassium), robbing trees of nutrients.

Toxic metals such as lead, zinc, copper, chromium, and aluminum are deposited in the forest by the atmosphere. When acid rain interacts with these metals, it stunts tree and plant growth, along with mosses, algae, nitrogen-fixing bacteria, and fungi.

Treating Acid Rain Deposition

A number of methods reduce acid deposition problems (e.g., liming) and aid in normalizing pH. Large amounts of hydrated lime or soda ash, added to lake waters, raise alkalinity and pH. However, some lakes are unreachable, too large and costly to treat, or have a high flow rate and become acidic soon after liming. The best way to slow or stop acid deposition is to limit chemical emissions at their source.

Legislation

In some countries, regulations limit atmospheric sulfur and nitrogen oxide emissions. Acidic pollutants from industry are reduced by (1) switching to fuels that have zero or low sulfur content, and (2) using smokestack scrubbers to reduce sulfur dioxide released. Requiring catalytic converters on vehicles limits these emissions from automobiles and trucks.

In 1991, Canada and the United States established the *Air Quality Accord,* which controls cross-border air pollution. This agreement established a permanent limit on acid emissions (i.e., 13.3 million tons in the United States and 3.2 million tons in Canada).

Oil Slicks, Radioactivity, and Thermal Problems

Large oil spills (e.g., the Exxon *Valdez*) cause tons of pollution along shorelines. One estimate suggests that one ton of oil is spilled for every million tons of transported oil.

Oil pollution is devastating to coastal wildlife, since even small oil amounts spread quickly across long distances to form deadly oil slicks. Once spilled, oil is hard to remove or contain, washing up along miles of shoreline. Efforts to chemically treat or sink spilled oil often disrupt marine ecosystems even more than the original spill.

Radioactivity

Large amounts of radioactive waste materials are created by nuclear power plants, industry, and mining. Dust and rock from uranium mining and refining contain radioactive contaminants, which cause problems due to runoff from mining sites.

Medical *radioactive tracers,* which include phosphorus (^{32}P), iron (^{59}Fe), and iodine (^{131}I), are used to detect and treat early-stage diseases (e.g., thyroid, breast cancer). Thallium (^{201}Tl) is used to detect heart disease since it binds tightly to well-oxygenated heart muscle. However, no matter how helpful they are, radioactive elements can enter sewage if not disposed of properly.

Thermal Pollution

Heat or *thermal pollution* has far-reaching and damaging ecological effects. It pollutes water by impacting aquatic organisms and animal populations.

> The release of a liquid or gas, which increases heat in a surrounding area, is known as *thermal pollution.*

Water temperature controls metabolic and reproductive activities in aquatic life. Most marine organisms are cold-blooded, and their body temperatures are controlled by the water temperature around them. Cold-blooded organisms have adapted to specific temperature ranges. If water temperatures change too much, metabolic processes break down. Unlike humans, who can adapt to wide temperature ranges, most organisms live in narrow temperature niches. When these niches change, marine organisms die.

Industries (e.g., electric power plants, refineries, metal smelters, paper mills, food processing, and chemical manufacturing) produce thermal pollution and often release high-temperature wastewater directly into rivers and lakes. This heated water disrupts ecosystems at the discharge site and downstream.

Noise Pollution

Noise pollution seems more subjective than other pollution types, since what affects one person may not bother someone else, but very loud noises like jack hammers and jet engines actually damage the inner ear. The Environmental Protection Agency created the *Noise Control Act* of 1972 to set limits on major sources of noise (e.g., construction, equipment and vehicles).

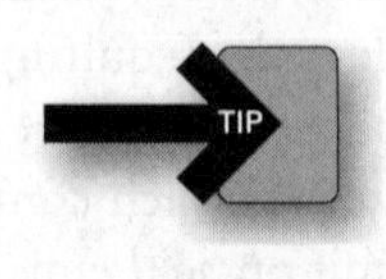

In the United States, the *Occupational Safety and Health Agency* (OSHA) regulates workplace noise pollution and safety. It sets limits for worker exposure to noise, as well as requiring protective safety equipment.

Looking to the Future

Science provides modern methods to reduce and treat pollutants before they enter the environment. However, it's important to consider how our activities affect the environment and what can be done to minimize the impact. Table 18.2 lists the many sources and types of pollution to monitor.

Table 18.2 There are many different sources of pollution.

POLLUTION SOURCE	COMPOSITION OF POLLUTANT
Automobiles	Burning of oil and gas produces carbon monoxide, VOCs, hydrocarbons, nitrogen oxides, peroxyacetyl nitrate, benzene, and lead.
Utility power plants	Burning of coal, oil, and gas produces nitrogen oxides, heavy metals, sulfur dioxide, and particulates.
Industry	Particulates, sulfur dioxide, nitrogen oxides, heavy metals, fluoride, CFCs, and dioxins are emitted by smoke stacks.
Incineration	Carbon monoxide, nitrogen oxides, particulates, dioxins, and heavy metals are produced by burning.
Biomass burning	Burning of grasslands, crop stubble, agricultural waste, organic fuel, and forests produces sulfur, methane, radon, carbon dioxide, nitrogen oxides, carbon monoxides, and particulates.
Small engines	Mowers, blowers, trimmers, chain saws, and other machines produce nitrogen oxides and hydrocarbons.
Disasters	Radiation leaks, chemical leaks, and burning of oil wells produce radiation, nitrogen oxides, carbon monoxide, sulfur, heavy metals, and particulates.
Mining	Rock breakdown and processing produces nitrogen oxides, heavy metals, radiation particles, and particulates.
Erosion	Road work and farm work produce dust, particulates, dried pesticides, and fertilizers.
Indoor air pollution	Carpeting, cooking, and other indoor products and activities produce formaldehyde, lead and asbestos dust, radon, and other incorporated chemicals.
Nature	Volcanic eruptions and forest fires produce dust and particles, sulfur dioxide, carbon monoxide, carbon dioxide, chlorine, nitrogen oxides, heavy metals, radon, and particulates.

Green Taxes

Income, land, and employment are taxed, while polluting processes have not historically been heavily taxed. However, policy makers are starting to set higher taxes on polluting activities and industries. *Green taxes* are gaining support, along with carbon taxes on fossil fuels; mining, energy, and forestry taxes; fishing and hunting licensing; garbage taxes; and effluent, emissions, and hazardous waste fees. These new and/or higher taxes are meant to clean up pollution problems and change behavior.

Since *Earth Day* was first established by President Richard Nixon on April 22, 1970, the public has awakened to resource limitations, endangered species, air and water pollutants, and environmental protection.

› Review Questions

Multiple-Choice Questions

1. The atmospheric layer largely responsible for absorbing the sun's ultraviolet (UV) radiation is the
   (A) thermosphere
   (B) cumulus cloud
   (C) troposphere
   (D) stratonimbus
   (E) ozone

2. Precipitation is considered acidic (e.g., rain, snow, sleet, hail, fog) if it has a pH less than
   (A) pH 5.6
   (B) pH 6.2
   (C) pH 7.0
   (D) pH 8.2
   (E) pH 9.0

3. The four main polluting contaminant types include all the following except
   (A) inorganic
   (B) organic
   (C) acid-base
   (D) drought
   (E) radioactive

4. Atmospheric gases blanketing the Earth exist in a mixture. What percent of this mixture is nitrogen (by volume)?
   (A) 8%
   (B) 20%
   (C) 36%
   (D) 58%
   (E) 80%

5. Taxes on fossil fuels, mining, energy, forestry, fishing and hunting licensing, garbage, effluent and emissions, and hazardous wastes are known as
   (A) recreation taxes
   (B) park taxes
   (C) green taxes
   (D) incentive taxes
   (E) single use taxes

6. The atmospheric layer where all the local temperature, pressure, wind, and precipitation changes take place is the
   (A) stratosphere
   (B) ionosphere
   (C) mesosphere
   (D) troposphere
   (E) thermosphere

7. Why is thermal pollution a problem for marine organisms?
   (A) They are hot-blooded and overheat.
   (B) Their metabolic processes break down.
   (C) They are adapted to wide temperature ranges.
   (D) They don't have a problem with it.
   (E) They live in many temperature niches.

8. When acid rain falls on limestone statues, monuments, and gravestones, discoloring and disfiguring surfaces, the process is known as
   (A) dispensation
   (B) sedimentation
   (C) dissolution
   (D) suspension
   (E) cementation

9. Brown urban smog is not emitted directly from specific sources, but formed in the atmosphere from nitrogen oxides and
   (A) inorganic compounds
   (B) volatile organic compounds
   (C) potassium chloride
   (D) fertilizer
   (E) helium

10. In the workplace, bad air mixed with mold spores has led to
   (A) shorter coffee breaks
   (B) increased productivity
   (C) sick building syndrome
   (D) reduced medical costs
   (E) greater appreciation of weekends

11. Looking at Figure 18.1, what dose would be lethal to 25% of the population?

(A) 4 ppm
(B) 7 ppm
(C) 9 ppm
(D) 11 ppm
(E) 14 ppm

12. City heat islands cause

(A) pollutants to collect
(B) residents to seek winter vacations
(C) less dust and lightning strikes
(D) reduced rainfall
(E) greater visibility

13. Total organic carbon (TOC) levels are used by hydrologists to check the health of freshwater as it affects biogeochemical processes and

(A) climate change
(B) nutrient cycling
(C) annual rainfall
(D) biological unavailability
(E) carbon nanotube levels

14. The Comprehensive Environmental Response Compensation and Liability Act (CERCLA) is commonly called the

(A) Superfund
(B) Clean Air Interstate Rule
(C) Liability Limitation Act
(D) Clean Water Act
(E) CAFE standards

15. The amount of dissolved oxygen in water depends on

(A) temperature
(B) water flow volume
(C) water flow velocity
(D) number of organisms using oxygen for respiration
(E) all of the above

16. Turbidity is a measure of water's

(A) transparency
(B) cloudiness
(C) chlorination
(D) coagulation
(E) flocculation

17. The amount of oxygen depletion in water depends on the speed at which a stream can take up atmospheric oxygen and replenish its

(A) color
(B) pathogens
(C) reaeration capacity
(D) turbidity
(E) minerals

18. The least likely way microscopic pathogens (e.g., bacteria, viruses, and protozoa) enter waterways is through

(A) untreated urban sewage
(B) farm runoff
(C) bilge water
(D) mining runoff
(E) family pet waste

› Answers and Explanations

1. C—The atmospheric layer closest to the Earth where life is protected from harmful cosmic radiation showers is the troposphere.

2. A—Natural rain has a pH of 5.6, so anything with a pH below that is considered acid rain.

3. D—Drought is a climatic condition not a contaminant.

4. E—Nitrogen makes up the largest component of the atmosphere.

5. C

6. D—The troposphere is the most active of the atmospheric layers.

7. B—Unlike humans, who can adapt to wide temperature ranges, most organisms live in narrow temperature niches and their metabolism breaks down at higher temperatures.

8. C

9. B—Urban smog, regulated by the EPA, is not emitted directly from specific sources, but formed in the atmosphere from nitrogen oxides and volatile organic compounds.

10. C—People suffer headaches, allergies, fatigue, nausea, and respiratory problems leading to greater medical costs, sick days off, and low productivity.

11. D—Starting at 25% of the population killed (vertical axis), intersect the dose response curve and then read the dose level (horizontal axis).

12. A—City heat islands collect pollutants such as dust and particulates.

13. B—Total organic carbon (TOC) levels are used by hydrologists to check the health of freshwater as it affects biogeochemical processes, bioavailability, and nutrient cycling.

14. A

15. E

16. B

17. C—Fast-flowing streams reoxygenate quickly, while deep, slow-flowing rivers take up oxygen much more slowly.

18. D—Mining runoff is often the source of heavy metal and chemical contaminants rather than pathogens.

Free-Response Questions

1. Sometimes when trying to help the environment, we create other problems. For example, methyl tertiary-butyl ether (MTBE), a by-product of natural gas, increases octane levels and burns cleaner than gasoline. In 1990, the Clean Air Act mandated MTBE be added to gasoline in areas with ozone problems. Unfortunately, MTBE was a serious groundwater pollutant. Like benzene, the most hazardous gasoline-related groundwater pollutant in the United States, MTBE (less toxic) changes groundwater color and causes it to smell and taste like turpentine in the smallest quantities. Because of this, California and 11 other states are phasing MTBE out and looking for alternatives to reduce pollution and fossil fuel emissions.

(a) Provide examples and descriptions of the three main water pollutants.

(b) Long-term environmental effects from the use of MTBE must be prevented by cessation of MTBE as a fuel additive. Groundwater changes that occur with MTBE-related reactions foul water use for public consumption or recreational use as well as environmentally by marine species in the polluted area. Other fuel methods or chemicals to increase octane levels must be found.

2. The Federal Centers for Disease Control estimate that 82% of all Americans have the widely used insecticide Dursban (now banned) in their bodies. To limit the use of this chemical and others, control pests, and protect waterways from pollution, people can (1) use pesticides sparingly, (2) focus on early identification of pests, (3) use natural controls (e.g., ladybugs eat aphids), and (4) plant naturally resistant native plants.

(a) What are other practical ways in which we can keep our air and water clean?
(b) What other pollutants have an impact on human health? Why?

Free-Response Answers and Explanations

1.

a. Municipal: Sewage is an example of municipal water pollution. Greater population growth and heavily populated urban centers produce vast quantities of wastewater, which if untreated can find its way into the groundwater and other freshwater supplies.

Industrial: MTBE is an example of an industrial water pollutant. Produced on a mass scale and mandated for use in gasoline by the government, MTBE rendered sources of water in California hazardous to the environment and undrinkable.

Agricultural: Fertilizers and pesticides are major contributors to agricultural water pollution due to runoff and poor irrigation practices.

b. Individuals can get MTBE information from California and then check MTBE use in their own state. With this information in hand, they can approach their state representatives about discontinuing the additive.

2.

a. We can keep our air and water clean by encouraging the food industry to use recycled packaging and natural dyes where possible and to keep toxic dyes out of landfills and groundwater. Walking or biking, instead of driving everywhere, also cuts down on acid, hydrocarbon, and nitrogen oxide emissions to the atmosphere and therefore to worldwide freshwater supplies.

b. Oil spills, acid rain, radioactivity, and noise pollution all impact human health. For example, very loud work environments can damage the inner ear and cause hearing loss.

› Rapid Review

- In general, four main categories of contaminants exist: organic, inorganic, radioactive, and acid-base.
- Green and carbon taxes are being assessed on fossil fuels, mining, energy, forestry, fishing and hunting licensing, garbage, effluent, emissions, and hazardous wastes.
- Thermal pollution, the release of liquid or gas that increases heat in a surrounding area, has far-reaching and damaging ecological effects by impacting aquatic organisms and animal populations.
- Cold-blooded organisms adapted to specific temperature ranges. If water temperatures change too much, metabolic processes break down. Unlike humans, who can adapt to wide temperature ranges, most organisms live in narrow temperature niches.
- In the workplace, sick building syndrome has caused workers to suffer headaches, allergies, fatigue, nausea, and respiratory problems.

- City heat islands collect pollutants so that areas downwind have much less visibility and greater rainfall due to condensation than their neighbors with cleaner air.
- Total organic carbon (TOC) levels are used by hydrologists to check the health of freshwater as it affects biogeochemical processes, nutrient cycling, biological availability, chemical transport and interactions.
- The amount of dissolved oxygen in water depends on temperature (the colder the water, the more oxygen that can be dissolved), water flow volume and velocity, and the number of organisms using oxygen for respiration.
- Microscopic pathogens (e.g., bacteria, viruses, and protozoa) enter waterways through untreated urban sewage, storm drains, pet waste, septic tanks, farm runoff, and bilge water causing sickness and/or death.
- Water turbidity, caused by suspended matter such as clay, silt, and organic matter, can also result from microscopic organisms blocking light through water and provide a medium for their growth.
- Ozone oxidation is a good water disinfectant method, but unlike chlorine, ozone doesn't stay in water after it leaves the treatment plant.
- Ultraviolet light is used to treat wastewater by killing microorganisms, but like oxidation, it is a one-time treatment at the plant.
- During coagulation, aluminum sulfate is added to raw sewage water, forming sticky blobs that snag small particles of bacteria, silt, and other impurities.

Global Change and Economics

IN THIS CHAPTER

Summary: Greenhouse gases are measurable. Global warming takes place when heat is retained near the Earth. The Earth is a living, recycling, and changing system.

Keywords

✪ Ozone, greenhouse gases, global warming, aerosols, carbon dioxide, nitrous oxide, methane, volatile organic compounds, carbon sequestration

---

Greenhouse Effect

Greenhouses, sometimes called hothouses, work by trapping the sun's heat. Their glass sides and roof let sunlight in, but keep heat from escaping, like a car parked in the sun with the windows rolled up. Similarly, greenhouses offer plants (e.g., tropical orchids) warm, humid environments, even when outside weather is dry, windy, or cold.

In Chapter 6, we saw how the atmosphere surrounds our planet like a blanket. It protects us from harmful cosmic radiation, regulates temperature and humidity, and controls the weather. The atmosphere is critical to life on this planet and provides the air we breathe.

The *greenhouse effect* describes how atmospheric gases prevent heat from being released back into space, allowing it to build up in the Earth's atmosphere.

The greenhouse effect keeps the Earth warm enough for living things to survive. However, if it gets too strong, it can overheat the planet. The problem is that even a few degrees higher creates problems for people, plants, and animals.

Greenhouse Gases

Greenhouse gases are a natural part of the atmosphere. They trap the sun's heat and preserve the Earth's surface temperature at a level needed to support life. Sunlight enters the atmosphere, passing through greenhouse gases like a lens. When it reaches the Earth's surface, land, water, and the biosphere absorb the sun's energy. Some heat is reflected back into space, but a lot stays in the atmosphere, heating the Earth. Figure 19.1 shows how these greenhouse gases trap energy in the atmosphere.

Greenhouse gases include water vapor, carbon dioxide, methane, nitrous oxide, halogenated fluorocarbons, ozone, perfluorinated carbons, and hydrofluorocarbons. Water vapor is the most important greenhouse gas, and human activity doesn't have much direct impact on its natural atmospheric level.

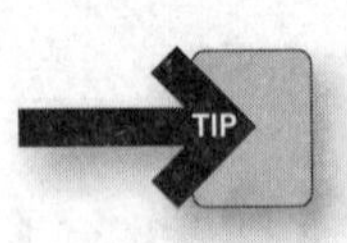

Global warming is caused by an increase in the levels of these gases. The greatest impact on the greenhouse effect has come from industrialization, which has increased the amounts of carbon dioxide, methane, and nitrous oxide in the atmosphere. Land clearing and fossil fuel burning have raised atmospheric concentrations of soot and other *aerosols* (air particles) as well.

According to climatic models (programmed with the volume of gases released into the atmosphere yearly), the planet is warming at a steep rate. Consequences of global

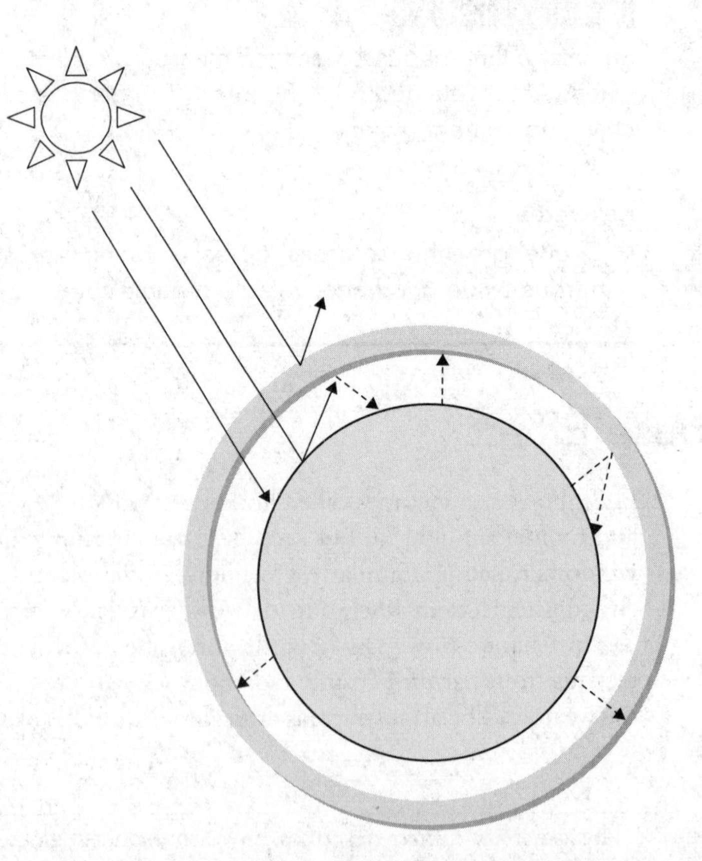

Figure 19.1 Greenhouse gases trap energy from the sun in the atmosphere, causing it to heat up.

warming (e.g., melting of polar ice and rising sea levels) are already taking place. Since 1991, the National Academy of Sciences has found clear evidence of global warming and recommends immediate greenhouse gas reductions. Depending on whether or not changes are made, temperature increases of between −16°C and −13°C in the next 100 to 200 years, with sea level increases of 1–8 meters are predicted.

Since greenhouse gases are long lasting, even if everyone stopped using their cars *today*, global warming would continue for another 150 years. Think of it like a speeding train; even when the engineer hits the brakes, the train's total speed and mass cause it to take a long time to stop. For this reason, it's important to speed up the use of alternative energy sources and curb and/or stop the release of greenhouse gases into the atmosphere.

Formation of Greenhouse Gases

Many industrial processes create greenhouse gases. When organic matter (e.g., table scraps, garden waste, and paper) is left in landfills, its decomposition forms methane and carbon dioxide. Sewage and water treatment plants also release these gases when breaking down wastes. Cement production, used in building roads and laying of building foundations, requires chemical processes that produce an assortment of greenhouse gases.

Carbon Dioxide

Carbon dioxide (CO_2) is a natural greenhouse gas and the biggest human-supplied contributor to the greenhouse effect (about 80%). A heavy, colorless gas, carbon dioxide is the main gas we exhale during breathing. It dissolves in water to form carbonic acid, is formed in animal respiration, and comes from the decay or combustion of plant and animal matter. Carbon dioxide is also used to carbonate drinks and is absorbed from the air by plants in photosynthesis.

The Earth's inhabitants don't have the option to stop breathing. However, the amount of carbon dioxide in the atmosphere is about 40% higher today than it was in the early 1800s. Figure 19.2 shows carbon dioxide concentration trends over the past 250 years.

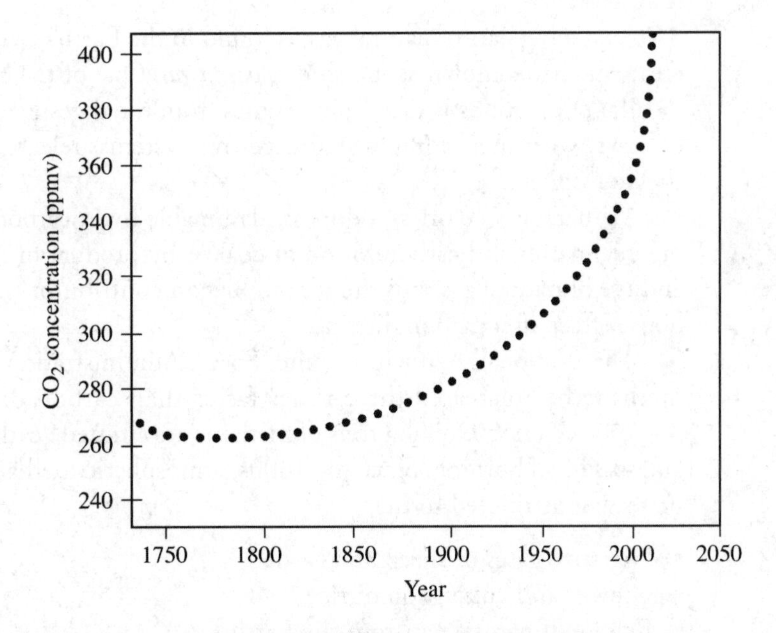

Figure 19.2 Carbon dioxide trends over the past 250 years.
Source: Oak Ridge National Laboratory (2002).

The industrial revolution is responsible for this jump. Ever since fossil fuels such as oil, coal, and natural gas were first burned to create energy for electricity and transportation fuel, carbon dioxide levels started to climb. Additionally, when farmers clear and burn weeds and crop stubble, carbon dioxide is produced.

Carbon dioxide gases also come from the Earth. When volcanoes explode, about 90–95% of the spewed gases are made up of water vapor and carbon dioxide.

Nitrogen Oxides

The colorless gas known as nitrous oxide is an atmospheric pollutant produced by combustion (e.g., 40% of total U.S. emissions in 2014), which traps heat much more efficiently than carbon dioxide. There are several ways nitrogen and oxygen team up in the atmosphere, including *nitrogen dioxide, nitric oxide,* and *nitrous oxide.* Since nitrogen oxides are stable gases and do not break down quickly, they build up in the atmosphere in greater and greater concentrations. In the sky, nitrogen dioxide creates a yellow-brown haze called smog.

Nitrogen combines with moisture in the atmosphere to form nitric acid. This comes down as rain and acidifies lakes and soils, killing fish and small animal populations and damaging forests. Acid particulates are precipitated, along with the leaching out of heavy metals, into water supplies. Scientists believe this increase comes from crop burning, industrial releases, and excess of nitrogen fertilizers used in agriculture.

Besides fossil fuel burning, nitrogen oxides are also produced by kerosene heaters, gas ranges and ovens, incinerators, and deforestation and leaf burning, as well as by aircraft engines and cigarettes. Lightning and natural soil sources also produce nitrogen oxides. Scientists estimate vehicles produce 40% of the pollution due to nitrogen oxide, with electric utilities and factories responsible for 50% of industrial emissions. The remaining 10% comes from other sources.

At high altitudes, nitrogen oxides are responsible for some ozone depletion. When ozone is thin or absent, the amount of solar ultraviolet radiation reaching the ground increases. This causes plant damage and injury to animals and humans in the form of skin cancers and other problems.

Methane

Methane, a big part of natural gas, is found in the Earth's crust. Underwater decaying plants create methane known as marsh or *swamp gas.* One of the best known sources of methane in rural populations is that which comes from the digestive tract of farm animals. Millions of cows, with their complicated digestive systems, release large amounts of methane to the atmosphere.

Methane is a colorless, odorless, flammable hydrocarbon released by the breakdown of organic matter and carbonization of coal. A by-product of the production, transportation, and use of natural gas, it is the second biggest contributor (i.e., 35% in 2007) to the greenhouse effect after carbon dioxide.

The National Aeronautics and Space Administration (NASA) reports atmospheric methane has increased three times faster than carbon dioxide and tripled in the past 30 years. Over 500 million tons of methane are emitted yearly from bacterial decomposition and fossil fuel burning. Since the 1800s, atmospheric methane levels have risen 145%. This increase is attributed to the

- Digestive gases of sheep and cattle
- Growth and cultivation of rice
- Release of natural gas from the Earth
- Decomposition of garbage and landfill waste

In the atmosphere, naturally occurring hydroxyl radicals combine with and remove methane, but as hydroxyl concentrations drop and methane emissions rise, overall methane concentrations will get higher.

Burning fossil fuels for energy creates greenhouse gases. When oil, gas, or coal burns, carbon in the fuel mixes with atmospheric oxygen to form carbon dioxide. Methane is produced from coal mining and certain natural gas pipelines. Rice production in paddy fields generates methane under water.

Enhanced Greenhouse Effect

Solar cycles and changes in the sun's radiation affect local climate and allow the sun's energy to reach the Earth's surface, keeping heat from escaping. The Earth gets slowly hotter. Industrial activity produces greenhouse gases, which serve as additional blankets to heat the Earth even more.

> The *enhanced greenhouse effect* caused by the burning of fossil fuels (oil, coal, and natural gas) increases global warming and changes the environment.

Global warming effects differ around the world and make it hard to predict exactly how the climate may change. Temperature increases are expected to be higher in polar areas than around the equator. Land temperatures might be higher than those over oceans. Rainfall might be heavier in some areas and lower in others.

TIP

Major climatic change greatly affects local weather through the frequency and intensity of storms. Some scientists fear the high number of major Atlantic hurricanes in recent years may be the beginning of severe climate change. Ranching, crops, pests, diseases, ocean levels, and native plant and animal populations would all be impacted.

Increasing global warming has also motivated people in different cities, states, industries, and countries to step up their work on developing "clean" energy options.

Carbon

Carbon is essential to life, but increasingly a problem as fossil fuel burning increases atmospheric carbon dioxide. However, unless you live in a polluted city or near industrial plants, it's hard to believe the Earth's immense atmospheric layers can't handle industrial pollutants. Global warming is often dismissed as a knee-jerk reaction of environmentalists.

In 2004, scientists reported in *Science* that the ocean had taken up nearly half of the carbon dioxide gas released into the air since the 1800s. This was good news, since less greenhouse gases means less global warming. However, oceans will reach their carbon dioxide limit.

Carbon uptake starts with plankton. These tiny organisms, drifting along on ocean currents, perform photosynthesis to produce energy and draw carbon out of the atmosphere. While building intricate calcium carbonate shells, they bind carbon as well. Eventually, these organisms die and sink to the bottom of the ocean dropping out of the carbon cycle. The rest dissolve in the low-calcium carbonate conditions of deep waters. Either way, their carbon doesn't return to the atmosphere for a very long time.

> *Carbon sequestration* removes and stores atmospheric carbon in carbon sinks (e.g., oceans, forests) through physical and biological processes like photosynthesis.

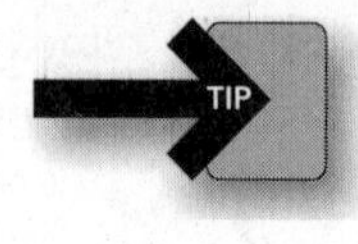

The amount of human-produced carbon dioxide being absorbed into the ocean is important. If carbon dioxide levels get high enough, the ocean's top layer will become more acidic, reducing calcium carbonate's availability to plankton. Then, as human-produced carbon dioxide sinks into the ocean, calcium carbonate dissolution may speed up.

This becomes especially significant when shell-making is impacted. In the last decade, scientists found that even small calcium carbonate decreases in seawater limited plankton's and coral's ability to build exoskeletons. If this continues over more centuries, organisms' ability to create shells may be compromised. If sinking shells from dead organisms dissolve in this shallower ocean water and their carbon returns into the atmosphere a lot sooner, the greenhouse effect will accelerate dramatically. The take home message is: if oceans become less effective as a sink for human-produced carbon dioxide, the buildup of atmospheric carbon dioxide will accelerate.

Greenhouse Gas Inventories

We've seen how the combustion of coal, oil, and natural gas cause global CO_2 levels to rise. Less well known is how trees absorb CO_2 during photosynthesis and then release it when they are cut down. The 34 million acres of tropical forests destroyed annually, about the size of New York state, release between 20–25% of total global CO_2 emissions.

In 1992, at the United Nations' Earth Summit in Rio de Janeiro, the international community first acknowledged the threat of climate instability. Over 185 nations agreed to reduce greenhouse gas emissions to their 1990 levels by 2012. More importantly, the participants agreed to stabilize atmospheric concentrations of greenhouse gases to prevent dangerous human interference with the global climate system. In June 2012, the "Rio+20" summit was held for nations to tackle unresolved environmental issues from the original Earth Summit conference in 1992 and plan future strategies.

The U.S. Environmental Protection Agency's Clean Air Markets Division developed the annual *Inventory of U.S. Greenhouse Gas Emissions and Sinks.* This EPA atmospheric inventory estimates, documents, and evaluates greenhouse gas emissions and sinks for all source categories. To update the report, the Inventory Program polls dozens of federal agencies, academic institutions, industry associations, consultants, and environmental organizations for up-to-date information. It also gets data from a network of continuous carbon dioxide emission monitors installed at most U.S. electric power plants.

In December 2008, the Department of Energy reported total U.S. greenhouse gas emissions of CO_2 levels rose by 17% from 1990 to 2007. If business-as-usual industrial output didn't change, global CO_2 levels would double by the end of the 21st century. In 2014, however, scientists were surprised to see continued decreases in CO_2 levels. The drop was thought to be due to cheaper, more available natural gas and a (>10%) decrease in coal burning.

Under the umbrella of the *Intergovernmental Panel on Climate Change* (*IPCC*), over 200 scientists and national experts worked together to develop guidelines to help countries create atmospheric inventories across international borders. Since then, scientists determined that stabilizing atmospheric levels of carbon dioxide will mean reducing CO_2 emissions and other heat-trapping gases to 80% of the 1990 global levels. According to some models, this means decreasing or stopping the release of 1.2 trillion tons of CO_2 by 2050. This isn't an impossible mission. Better energy efficiency throughout the global economy could prevent one-third of emissions while cutting energy costs.

Scientists use natural and industrial emission inventories as tools to develop atmospheric models. Policy makers and regulatory agencies use these inventories to check policy compliance and emission rates. Most inventories contain the following information:

- Chemical and physical identity and properties of pollutants
- Geographical area affected

- Time period when emissions were generated
- Types of activities that cause emissions
- Description of methods used
- Data collected

About 25% of the needed greenhouse gas decreases, about 370 billion tons, could be achieved by stopping tropical deforestation, restoring degraded lands, and improving land productivity worldwide through the use of best practices in agriculture and forestry. At local and regional levels, individuals have an important role.

International climate policy usually focuses on lowering emissions by adopting alternative energy options. Options without negative climate and biodiversity impacts (e.g., solar and wind power) are crucial to lowering energy-related emissions.

Environmental leaders and organizations are working with industries to reduce their carbon dioxide emissions, as well as enhance their competitive advantage. Industry has also turned to solar or wind power to cut costs, along with seeking new methods (e.g., nanotechnology) for energy transmission.

Ozone

Ozone (O_3) has a distinctive odor and blue color, while normal oxygen is odorless and colorless. Concentrated in a thin upper stratospheric layer, ozone is a very reactive form of oxygen. Constantly created and destroyed in the stratosphere, ozone levels are relatively constant and affected naturally by sunspots, the seasons, and latitude. There are also yearly and geographical drops in ozone levels, followed by a recovery.

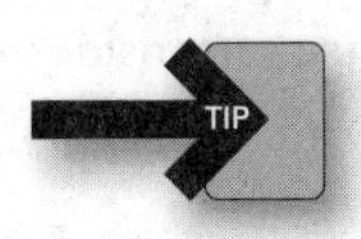

Ozone is an atmospheric bodyguard with a crucial role in protecting life on Earth. It is largely responsible for absorbing the sun's ultraviolet (UV) radiation. Most importantly, it absorbs the fraction of ultraviolet light called *UVB radiation.* Table 19.1 lists UV radiation types and effects.

Ultraviolet radiation is a bad, bad thing! It causes breaks in the body's nuclear proteins (DNA) leaving the door open for associated health problems. UVB is connected with skin cancer and cataracts. It is also harmful to crops, materials, and marine organisms. Atmospheric ozone blocks intense UV radiation from reaching the surface of Earth and the plants and animals living there. However, with an ever-increasing human population, atmospheric changes prevent ozone recovery.

Ozone Depletion

For the past 50 years, *chlorofluorocarbons* (CFCs) held the answer to many material problems. They were stable, nonflammable, not too toxic, and cheap to produce. They had a variety of uses including applications as refrigerants, solvents, and foam blowing agents. Chlorine formulations did everything from disinfect water to serve as solvents (e.g., methyl chloroform

Table 19.1 Ultraviolet radiation has many negative effects on living organisms.

UV RADIATION TYPE	WAVELENGTHS (NM)	EFFECTS ON LIFE
UVA	400–320	Fairly safe; tanning but not burning
UVB	320–290	Harmful; sunburn, skin cancer, other problems
UVC	290–200	Very harmful, but mostly absorbed by ozone

and carbon tetrachloride) in chemistry labs. Roughly 84% of stratospheric chlorine comes from human-made sources, while only 16% comes from natural sources. Unfortunately, these compounds don't just break down in the atmosphere and disappear. They linger and are carried by winds into the stratosphere to destroy ozone faster than it is naturally created.

CFCs break down only by exposure to strong UV radiation. When that happens, CFC releases chlorine, which damages the ozone layer. Scientists have found that 1 atom of chlorine can destroy over 100,000 ozone molecules.

In the 1950s, a "hole" or thinning of the ozone layer over Antarctica was first reported. This annual event of extremely low ozone levels showed drops of over 60% during bad years. Further research found ozone also thinned over North America, Europe, Asia, Australia, South America, and Africa. In fact, since the first reports, Earth observation satellites, such as the European Space Agency's *Envisat*, track the yearly arrival and shape of the ozone hole.

Scientific outcry over ozone depletion led to a 1978 ban on aerosol CFCs in several countries, including the United States. In 1985, the Vienna Convention was tasked to gather international cooperation in reducing CFC levels by half.

The good news is that atmospheric chlorine stopped rising in 1997–98, and if this trend continues, natural ozone recovery should occur in about 50 years.

The ozone layer will recover over time, but other problem gases (e.g., sulfur hexafluoride) are released during aluminum smelting, electricity production, magnesium processing, and semiconductor manufacturing and need to be addressed as well.

Nature's Part

Green plants use the sun's energy and carbon dioxide from the air for photosynthesis. This is a good thing, because they soak up carbon dioxide in the process. Plants are considered to be carbon dioxide storehouses. During the photosynthetic cycle, they form carbohydrates, which make up the foundation of the food chain.

Forests absorb carbon dioxide in a big way. We have learned how forests build up a significant supply of stored carbon in their tree trunks, roots, stems, and leaves. Then, when the land is cleared, this stored carbon is converted back to carbon dioxide by burning or decomposition.

The oceans are another big player in this process of carbon dioxide absorption. They also absorb carbon dioxide from the atmosphere and act as a moderating influence on temperature ranges.

Climate Change

Climate change includes temperature increases, rising sea levels, rainfall changes, and more extreme weather events. Scientific data discussed at the *2002 World Summit on Sustainable Development* and the *2004 United Nations' Convention on Biological Diversity* suggest global warming is causing shifts in species habitat and migrations by an average of 6.1 kilometers per decade toward the poles. This shift, predicted by climate change models, notes that spring arrives 2.3 days earlier per decade, on average, in temperate latitudes. Entire boreal and polar ecosystems are also showing the effects of global warming.

Climate change describes the difference in either the average state of the climate, or its variability, taking place over an extended period of time.

Past changes in global climate resulted in major shifts in species' ranges and a huge reorganization of biological communities, landscapes, and biomes during the last 1.8 million years. These changes took place in a much simpler world than today, and with little or no pressure from human activities. Species biodiversity is impacted by climate changes, as well as human pressure and adaptation.

Global warming is not just a theory anymore. Data confirm it is definitely happening, and we are our own biggest enemy. Environmental biologists and chemists calculated that in terms of equivalent units of carbon dioxide, humans are releasing roughly the same amount of greenhouse gases into the atmosphere as that of a Mount St. Helens eruption every two days. Not a good thing.

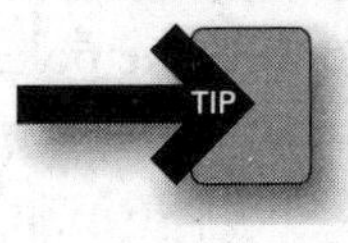

Global scientific agreement within the past 5 years is that the growing levels of heat-trapping gases are definitely affecting global climate and regional weather patterns. Those changes are causing a domino effect and impacting biodiversity.

Scientists have already seen regular and extensive effects on many species and ecosystems. In the past 30 years alone, climate change has resulted in large shifts in the distribution and abundance of many species. These impacts range from the disappearance of toads in Costa Rica's cloud forests to the death of coral reefs throughout the planet's tropical marine environments.

Global Economics

The World Bank was established in 1945 to aid war-torn Europe and Japan, but in the 1950s, its emphasis shifted to helping third-world countries. Unfortunately, many of its loans ended up being shortsighted. For example, in Botswana, a multimillion dollar loan was awarded to increase beef production for export, even though severe overgrazing on delicate grasslands was a known problem. This environmentally ill-considered project and others like it failed. Consequently, the United States and other concerned countries insisted future World Bank loans be subject to thorough environmental impact reviews before being granted. Only time will tell if the World Bank, global policy makers, or economists will work together for the environment.

› Review Questions

Multiple-Choice Questions

1. Greenhouses work

(A) by trapping the sun's heat
(B) when the walls and ceiling are painted green
(C) by cooling hot house plants to the dew point
(D) best when arctic mosses are grown
(E) when generators are used to maintain temperature

2. Natural and industrial emission inventories contain

(A) geographical data
(B) the time period when emissions were generated
(C) types of activities causing emissions
(D) the chemical and physical identity of pollutants
(E) all of the above

3. Greenhouse gases include all the following, except

(A) water vapor
(B) nitrous oxide
(C) carbon dioxide
(D) methane
(E) butyric acid

4. Climate change includes all the following except

(A) a rise in sea levels
(B) changes in rainfall patterns
(C) cooling of the Earth's core
(D) rising temperatures
(E) increased incidence of extreme weather

5. When fossil fuels are burned, adding to greenhouse gases and creating global warming, it is known as

(A) biodiversity lithification
(B) gasification
(C) deforestation
(D) dissolution
(E) enhanced greenhouse effect

6. Important decreases in greenhouse gas could be achieved by all the following except

(A) stopping tropical deforestation
(B) mining and burning more coal
(C) increasing agricultural and forestry best practices
(D) restoring and conserving degraded lands
(E) improving land productivity worldwide

7. The 2004 United Nations' Convention on Biological Diversity suggests that global warming is causing

(A) poor television reception
(B) an increase of mosquitoes in southern climates
(C) species habitat and migration shifts averaging 6.1 kilometers per decade toward the poles
(D) developing countries to use much more energy per person than developed countries.
(E) coastal areas to support increased numbers of mollusc species

8. Stabilizing atmospheric levels of carbon dioxide will mean reducing carbon dioxide emissions and other heat-trapping gases to what percent of 1990 global levels?

(A) 20%
(B) 35%
(C) 50%
(D) 80%
(E) 95%

9. Which of the following does not describe nitrogen oxides?

(A) They are stable gases.
(B) They are blue in color.
(C) They do not break down quickly.
(D) They build up in the atmosphere.
(E) They appear as a yellow-brown haze.

10. If the amount of human-produced carbon dioxide being absorbed into the oceans gets high enough, the ocean's top layer may become increasingly

(A) acidic
(B) basic
(C) opaque
(D) neutral
(E) murky

11. What organisms perform photosynthesis to produce energy and draw carbon out of the atmosphere?

(A) Whales
(B) Tulips
(C) Plankton
(D) Algae
(E) Deep sea tube worms

12. Ranching, crops, number of pests and diseases, and native plants and animals will be impacted by

(A) greater number of vegetarians
(B) new farming subsidies
(C) lower levels of carbon dioxide
(D) climate change
(E) weaker environmental policies

13. Removal and storage of atmospheric carbon in carbon sinks (e.g., oceans, forests) through physical and biological processes is called

(A) acid rain
(B) carbon sequestration
(C) lithification
(D) dissolution
(E) green energy

14. Compared to carbon dioxide, how do nitrogen oxides trap heat in the atmosphere?

(A) They don't trap heat at all.
(B) They trap heat about 1% less efficiently.
(C) They trap heat much more efficiently.
(D) They trap heat about 10% less efficiently.
(E) They trap heat much less efficiently.

15. What group developed guidelines to help countries create atmospheric inventories across international borders?

(A) EPA
(B) IPCC
(C) NASA
(D) CDC
(E) USDA

16. What percent of total greenhouse gases comes from carbon dioxide?

(A) 10%
(B) 30%
(C) 50%
(D) 70%
(E) 80%

17. Depending on the level of greenhouse gases in the next 100 to 200 years, scientists predict temperature increases between −16°C and −13°C, with sea levels rising up to

(A) 2 meters
(B) 5 meters
(C) 8 meters
(D) 9 meters
(E) 15 meters

› Answers and Explanations

1. A—Greenhouse gases trap the sun's heat and keep the Earth's surface temperature at a level needed to support life. Sunlight passes through these gases like a lens.

2. E—The Intergovernmental Panel on Climate Change developed a set of guidelines to help countries create atmospheric inventories across international borders.

3. E—Butyric acid is a building block of lipids (fats) and not a greenhouse gas.

4. C—The Earth's core has not cooled measurably in human history.

5. E—These gases are added to existing natural atmospheric gases.

6. B—Fossil fuel burning increases greenhouse gases as well as soot and aerosols.

7. C—Since many species have narrow temperature tolerances, they must move to cooler average temperature areas to survive.

8. D

9. B—They are colorless unless combined with other gases.

10. A—As carbon dioxide sinks into the ocean, it can reduce total calcium carbonate levels and cause greater acidity.

11. C—Drifting along on ocean currents, plankton perform photosynthesis, producing energy and drawing carbon out of the atmosphere.

12. D—Major climatic change greatly affects local weather through the frequency and intensity of storms.

13. B—While building intricate calcium carbonate shells, plankton bind further carbon.

14. C—Since nitrogen oxides are stable gases and don't break down quickly, they build up in the atmosphere.

15. B—The Intergovernmental Panel on Climate Change included 200 scientists and experts.

16. E

17. C—Large water volumes stored as ice are poured back into the oceans as ice melts.

Free-Response Questions

1. Oceanographers mapped the ocean's carbon chemistry. They compared what the ocean looked like before the industrial revolution (i.e., subtracting out carbon from fossil fuels) to current carbon values. Their findings, reported in the July 16, 2004, issue of *Science*, show that where human-produced carbon dioxide has sunk deep enough, the layer of carbonate-dissolving ocean water is now roughly 200 meters closer to the surface.
   (a) What effects can too much CO_2 have on the ocean's chemistry?
   (b) Why are policy makers looking for a way to pull carbon out of the atmosphere?

2. In a National Science Foundation Report on Global Warming, Anthony Leiserowitz at the University of Oregon Survey Research Laboratory asked Americans their opinions on global warming. Some of the survey results showed that
   - Of 92% of Americans who had heard of global warming, over 90% thought the United States should reduce its greenhouse gas emissions.
   - 77% support government regulation of carbon dioxide as a pollutant and investment in renewable energy (71%).
   - 76% want the United States to reduce greenhouse gas emissions regardless of what other countries do.

(a) Why is it important to raise public awareness about global warming?
(b) Should the population be aware of the risk to health if the hole in the ozone stops recovering yearly?

Free-Response Answers and Explanations

1.

a. Too much carbon dioxide dissolved into our oceans can acidify it, harming marine life. Overburdening any environment's capacity to manage levels of greenhouse gases places a stress on the ecosystem and biome, which can result in a loss of organisms and habitats. When carbon dioxide dissolves in water, it forms carbonic acid which releases hydrogen atoms. These hydrogen atoms combine with carbonate in seawater forming bicarbonate, which does not escape back into the atmosphere easily.

b. Carbon sequestration removes and stores atmospheric carbon in carbon sinks (e.g., oceans, forests) through physical and biological processes like photosynthesis. Policy makers want to find ways to sequester carbon until clean alternate energy sources can provide the bulk of the world's energy needs.

2.

a. Rather than instill anxiety or panic in a population, raising public awareness is an important step in finding a better way to scale back harmful human behavior. This is especially true in a country like the United States, which greatly contributes increased levels of carbon dioxide around the world. Raising public awareness is also a key step in holding policy makers, and society at large, to standards of accountability.

b. Ozone, an atmospheric bodyguard, has a crucial role in protecting life on Earth. It is largely responsible for absorbing the sun's harmful ultraviolet (UV) radiation. Without it, skin cancer and other health problems would become much more prevalent and affect large percentages of the world's population. People need to know their choices (i.e., chlorofluorocarbon use as a propellant in spray cans) have consequences, and each person can make a difference (e.g., using a pump sprayer) if they are informed.

❯ Rapid Review

- Greenhouse gases (e.g., roughly 79.1% nitrogen, 20.9% oxygen, 0.03% carbon dioxide, and trace amounts of others) are a natural part of the atmosphere.
- The term *greenhouse effect* describes how atmospheric gases prevent heat from being released back into space, allowing it to build up in the Earth's atmosphere.
- Greenhouse gases also include water vapor, methane, nitrous oxide, halogenated fluorocarbons, ozone, perfluorinated carbons, and hydrofluorocarbons.
- Water vapor is the most important greenhouse gas, but human activity doesn't have much direct impact on its natural atmospheric level.
- Global warming is caused by an increase in greenhouse gas levels.
- The greatest impact on the greenhouse effect has come from industrialization and increases in the amounts of carbon dioxide, methane, nitrous oxide, soot, and aerosols.
- Carbon dioxide (CO_2) is a natural greenhouse gas and the biggest human-supplied contributor to the greenhouse effect (about 70%).

- The 34 million acres of tropical forests destroyed annually release between 20–25% of total global CO_2 emissions.
- Nitrous oxide is a stable atmospheric pollutant produced by combustion.
- Nitrogen and atmospheric moisture form nitric acid, which comes down as acid rain and acidifies lakes and soils, kills fish and animals, and damages forests.
- Ozone, an atmospheric bodyguard, has a crucial role in protecting life on Earth by absorbing the sun's ultraviolet (UV) radiation.
- Ultraviolet radiation breaks down the body's DNA and causes skin cancer and cataracts.
- The enhanced greenhouse effect caused by the burning of oil, coal, and natural gas increases global warming and changes the environment.
- Drifting on ocean currents, plankton use photosynthesis to produce energy and draw carbon out of the atmosphere. While building intricate calcium carbonate shells, they bind carbon as well.
- Major climatic change greatly affects local weather through the frequency and intensity of storms.
- Carbon sequestration removes and stores atmospheric carbon in carbon sinks (e.g., oceans, forests) through physical and biological processes like photosynthesis.
- If oceans become less effective in serving as a sink for human-produced carbon dioxide, atmospheric carbon dioxide buildup will accelerate.
- Global warming is causing shifts in species habitat and migrations averaging 6.1 kilometers per decade toward the poles.
- The EPA atmospheric inventory estimates, documents, and evaluates greenhouse gas emissions and sinks for all source categories.
- If business-as-usual industrial output doesn't change, global CO_2 levels will double by the end of the 21st century.
- Roughly 84% of stratospheric chlorine comes from human-made sources, while only 16% comes from natural sources.
- Ozone reduction allows harmful UV radiation through the atmosphere, causing skin cancer, eye damage, and other harmful effects.

Build Your Test-Taking Confidence

Sample Exam 1—Multiple-Choice Questions

ANSWER SHEET

1 A B C D E
2 A B C D E
3 A B C D E
4 A B C D E
5 A B C D E
6 A B C D E
7 A B C D E
8 A B C D E
9 A B C D E
10 A B C D E
11 A B C D E
12 A B C D E
13 A B C D E
14 A B C D E
15 A B C D E
16 A B C D E
17 A B C D E
18 A B C D E
19 A B C D E
20 A B C D E
21 A B C D E
22 A B C D E
23 A B C D E
24 A B C D E
25 A B C D E
26 A B C D E
27 A B C D E
28 A B C D E
29 A B C D E
30 A B C D E
31 A B C D E
32 A B C D E
33 A B C D E
34 A B C D E
35 A B C D E
36 A B C D E
37 A B C D E
38 A B C D E
39 A B C D E
40 A B C D E
41 A B C D E
42 A B C D E
43 A B C D E
44 A B C D E
45 A B C D E
46 A B C D E
47 A B C D E
48 A B C D E
49 A B C D E
50 A B C D E
51 A B C D E
52 A B C D E
53 A B C D E
54 A B C D E
55 A B C D E
56 A B C D E
57 A B C D E
58 A B C D E
59 A B C D E
60 A B C D E
61 A B C D E
62 A B C D E
63 A B C D E
64 A B C D E
65 A B C D E
66 A B C D E
67 A B C D E
68 A B C D E
69 A B C D E
70 A B C D E
71 A B C D E
72 A B C D E
73 A B C D E
74 A B C D E
75 A B C D E
76 A B C D E
77 A B C D E
78 A B C D E
79 A B C D E
80 A B C D E
81 A B C D E
82 A B C D E
83 A B C D E
84 A B C D E
85 A B C D E
86 A B C D E
87 A B C D E
88 A B C D E
89 A B C D E
90 A B C D E
91 A B C D E
92 A B C D E
93 A B C D E
94 A B C D E
95 A B C D E
96 A B C D E
97 A B C D E
98 A B C D E
99 A B C D E
100 A B C D E

Sample Exam 1—Multiple-Choice Questions

(Time—1 hour and 30 minutes)

1. The amount of open space between soil particles is called

(A) aridity
(B) saturation
(C) crystallization
(D) porosity
(E) lithification

2. Carbon dioxide, methane, and nitrous oxide are all

(A) used in carbonated drinks
(B) used by dentists
(C) greenhouse gases
(D) produced by gaseous cattle
(E) inorganic gases

3. The Earth's water is stored for long periods of time in

(A) swimming pools
(B) caverns
(C) mountain streams
(D) stock ponds
(E) reservoirs

4. CAFE Tier 2 standards reduced nitrous oxide emissions and

(A) calcium emissions
(B) methane emissions
(C) ozone emissions
(D) CFC emissions
(E) sulfur emissions

5. Greenhouse gases are

(A) mostly released from gardening sheds
(B) not a growing problem
(C) primarily argon and xenon
(D) a natural part of the atmosphere
(E) always found in humid forests

6. The measurement of the number of hydrogen ions in water is called

(A) pH
(B) hydrogenation
(C) total carbon content
(D) dissolved oxygen
(E) radioactive decay

7. Pitchblende is another name for

(A) calcium chloride
(B) iodine
(C) bismuth
(D) uranium oxide
(E) magnesium sulfate

8. The biggest human-supplied gas to the greenhouse effect is

(A) carbon dioxide
(B) sulfur dioxide
(C) ammonia
(D) methane
(E) potassium chloride

9. What percent of the world's population lives within 60 km of a coastline?

(A) 15%
(B) 30%
(C) 50%
(D) 66%
(E) 72%

10. Since heat moves from hotter to colder areas, the Earth's heat moves from its fiery center toward the

(A) North Pole
(B) oceans
(C) forests
(D) core
(E) surface

11. The average amount of time an element, like carbon or calcium, spends in a geological reservoir is known as

(A) pool time
(B) cosmic time
(C) dissolution time
(D) standard time
(E) residence time

12. Water held within plants returns to the atmosphere as a vapor through a process called

(A) evacuation
(B) transpiration
(C) condensation
(D) resuscitation
(E) sedimentation

13. Infectious, pathological, and chemotherapy waste are types of

(A) fossil fuel waste
(B) agricultural waste
(C) biological waste
(D) radioactive waste
(E) cyber waste

14. Where fresh river water joins salty ocean water, it is known as

(A) polluted
(B) pure
(C) silted
(D) brackish
(E) backwash

15. Reykjavik, Iceland's capital, is

(A) sooty and polluted
(B) an ice-covered wasteland
(C) balmy and temperate
(D) extremely difficult to reach by train
(E) heated completely by geothermal energy from volcanic sources

16. What book, written by marine biologist Rachel Carson in 1962, warned of herbicide and pesticide hazards?

(A) *The Polluted Spring*
(B) *The Quiet Stream*
(C) *The Killing Spring*
(D) *The Silent Spring*
(E) *The Flooding Stream*

17. Atmospheric gases blanketing the Earth exist in a mixture. What percent of this mixture is oxygen (by volume)?

(A) 8%
(B) 20%
(C) 36%
(D) 58%
(E) 79%

18. Africa, Antarctica, Australia, Eurasia, North America, and South America make up the

(A) hot continents
(B) prime vacation spots for families
(C) six major land masses
(D) six wonders of the world
(E) highest producers of methane globally

19. Nitrogen's residence time in the environment is approximately

(A) 4 years
(B) 40 years
(C) 4,000 years
(D) 4 million years
(E) 400 million years

20. Hazardous waste is all the following except

(A) highly flammable
(B) corrosive
(C) a health hazard
(D) nontoxic
(E) reactive

21. The 1979 Hollywood movie *The China Syndrome* was about

(A) nuclear power
(B) hydroelectric power
(C) wind power
(D) solar power
(E) geothermal power

22. The law of horizontality describes

(A) igneous rock formation
(B) how older rock layers are deeper underground than younger layers
(C) art deco design
(D) volcanic blast patterns
(E) younger rock layers deeper than older layers

23. When water soaks into the ground, it's called

(A) evaporation
(B) transpiration
(C) infiltration
(D) respiration
(E) precipitation

24. What simple mode of power generation was once used to grind grain and pump water?

(A) Hydroelectric
(B) Nuclear
(C) Dog sled
(D) Electric generator
(E) Wind

25. To conserve water, major U.S. cities require new building construction to use

(A) foreign plywood
(B) water-saving toilets
(C) copper wiring
(D) slow-running faucets
(E) small-head shower nozzles

26. Approximately how many people around the world depend on fuel wood as their main fuel source?

(A) 500,000
(B) 800,000
(C) 1.5 billion
(D) 3 billion
(E) 4 billion

27. The layer of rock that drifts slowly over the supporting, malleable, upper mantle layer is called

(A) a geological plate
(B) a guyot
(C) magmic rock
(D) tellurite
(E) bedrock

28. What is thought to create the most greenhouse gases?

(A) Belching cows
(B) Jet contrails
(C) Fertilizers
(D) Motorcycles
(E) Burning of fossil fuels

29. While 16% of stratospheric chlorine comes from natural sources, what percentage comes from human-made sources?

(A) 23%
(B) 37%
(C) 58%
(D) 84%
(E) 90%

30. When rain has no time to evaporate, transpire, or move into groundwater reserves, it becomes

(A) no problem
(B) runoff
(C) infiltration
(D) permeation
(E) wetlands

31. When plants absorb carbon dioxide and sunlight to make glucose and build cellular structures, it is part of the

(A) biological carbon cycle
(B) astrological cycle
(C) calcium cycle
(D) remediation cycle
(E) sulfur cycle

32. On a per capita basis, which country is the world's largest nuclear power producer?

(A) Argentina
(B) Spain
(C) India
(D) France
(E) Canada

33. Pollutants from agriculture, storm runoff, lawn fertilizers, and sewer overflows are

(A) point source
(B) open source
(C) nonpoint source
(D) easy to trace
(E) always absorbed by wetlands

34. The middle layer between the lower stratosphere and the thermosphere is called the

(A) troposphere
(B) ionosphere
(C) biosphere
(D) outer space
(E) mesosphere

35. Which process removes and stores atmospheric carbon in carbon sinks (e.g., oceans, forests) through physical and biological processes like photosynthesis?

(A) Carbon sequestration
(B) Denudation
(C) Global warming
(D) Temperature inversion
(E) Volcanism

36. An air current found between 10–14 km above the Earth's surface in the troposphere is the

(A) hurricane
(B) Coriolis effect
(C) equatorial wind
(D) jet stream
(E) typhoon

37. Which gas is the second biggest additive to the greenhouse effect at around 20%?

(A) Propane
(B) Methane
(C) Carbon dioxide
(D) Neon
(E) Helium

38. When flowing water creates electricity, it is called

(A) rafting power
(B) hydroelectric power
(C) solar power
(D) dispersion power
(E) geothermal power

39. All the following are water pollution contaminants except

(A) heat
(B) fertilizers
(C) Coho salmon
(D) oil spills
(E) radioactivity

40. Tidal turbines driven by ocean currents are called

(A) mills
(B) funnels
(C) PV cells
(D) dams
(E) barrages

41. Which of the following is a layer of the atmosphere?

(A) Decosphere
(B) Cryosphere
(C) Stratosphere
(D) Lithosphere
(E) Blue sphere

42. When elemental isotopes are spun off and energy is emitted in the form of alpha, beta, and gamma particles, it is known as

(A) a solar flare
(B) fireworks
(C) chemical erosion
(D) halogens
(E) radioactive decay

43. How many oxygen atoms does ozone have?

(A) 1
(B) 2
(C) 3
(D) 4
(E) 5

44. What was used to light homes before the invention of the light bulb in 1879?

(A) Kerosene
(B) Propane
(C) Wind power
(D) Creosote
(E) Sheep oil

45. Point source and nonpoint source pollutants are important in which major environmental cycle?

(A) Calcium cycle
(B) Carbon cycle
(C) Hydrology cycle
(D) Rock cycle
(E) Sulfur cycle

46. Made of semiconducting materials, which one of these converts sunlight directly into electricity?

(A) Mood rings
(B) Hydroelectric power plants
(C) Nuclear fuel rods
(D) Photovoltaic cells
(E) Magnifying lens

47. Which of the following absorbs the greenhouse gas, carbon dioxide, in a big way?

(A) Streams
(B) Forests
(C) Uranium oxide
(D) Cattle
(E) Champagne

48. The annual thinning of the ozone layer (i.e., extremely low ozone levels) shows drops of over

(A) 15%
(B) 30%
(C) 60%
(D) 75%
(E) 90%

49. Most streams follow a branching drainage pattern, which is known as

(A) dendritic drainage
(B) radial drainage
(C) rectangular drainage
(D) composite drainage
(E) fractionated drainage

50. What are severe storms called in the Northwestern Pacific Ocean and Philippines?

(A) Tsunamis
(B) Blue northern
(C) Rainy season
(D) Typhoons
(E) Hurricanes

51. When harnessing wind power, propeller-like turbine blades are used to generate

(A) tourism
(B) electricity
(C) oxygen
(D) heat
(E) coastal winds

52. The Aurora Borealis and Aurora Australis are found in which atmospheric layer?

(A) Mesosphere
(B) Stratosphere
(C) Tropopause
(D) Thermosphere
(E) Troposphere

53. Getting more oil from a previously drilled deposit using other methods is called a

(A) risky endeavor
(B) barrage
(C) secondary recovery
(D) passive extraction
(E) tectonic convergence

54. Acid rain has been measured in

(A) the United States
(B) the Netherlands
(C) Australia
(D) Japan
(E) all of the above

55. Radioactive waste material, created by mining, includes

(A) boat sewage
(B) uranium dust
(C) oil spills
(D) biomedical waste
(E) tube worms

56. Which of the following public reactions was not a result of the first Earth Day?

(A) Knowledge of endangered species
(B) Awakening to limited resources
(C) Impression that fossil fuels would last indefinitely
(D) Protection of the planet
(E) Importance of air and water pollutants

57. Rainfall and streams on the east side of the Rocky Mountains drain to the Atlantic Ocean, while flowing water from the Rocky Mountain's western slopes runs to the

(A) Gulf of Mexico
(B) Hudson Bay
(C) Pacific Ocean
(D) Indian Ocean
(E) Great Lakes

58. Global warming works a lot like

(A) an island hotspot
(B) a greenhouse
(C) the Gulf Stream
(D) a guesthouse
(E) an electric blanket

59. Of the 5,513 known species of mammals, what percent is currently vulnerable or endangered?

(A) 2%
(B) 7%
(C) 10%
(D) 17%
(E) 25%

60. Approximately how many streams has the U.S. Geological Survey (USGS) counted in the United States?

(A) 100,000 to 200,000
(B) 550,000 to 1 million
(C) 1.5 to 2 million
(D) 2 to 3 million
(E) 3 to 4 million

61. What measures the rate of heat loss from exposed skin to the surrounding air?

(A) Thermometer
(B) Wind chill factor
(C) Barometer
(D) Wind shear
(E) Seismograph

62. At the upper edge of the zone of saturation and the lower edge of the zone of aeration is the

(A) bedrock
(B) lithosphere
(C) epicenter
(D) water table
(E) mantle

63. The Mississippi River drains 320 million km^2 of land area and is a

(A) first-order stream
(B) second-order stream
(C) fifth-order stream
(D) tenth-order stream
(E) twelfth-order stream

64. The Mariana Trench (five times as deep as the Grand Canyon) is over

(A) 1,000 meters deep
(B) 3,000 meters deep
(C) 5,000 meters deep
(D) 8,000 meters deep
(E) 11,000 meters deep

65. Which atmospheric layer can reach temperatures of nearly 2,000°C?

(A) Tropopause
(B) Mesosphere
(C) Thermosphere
(D) Troposphere
(E) Lithosphere

66. In 1998, the United Nations declared the

(A) first ban on clear-cutting of forests
(B) International Year of Science
(C) end to Daylight Savings Time
(D) International Year of the Ocean
(E) war on drug trafficking

67. A Class 4 hurricane has winds measuring

(A) 65 to 82 knots
(B) 83 to 95 knots
(C) 96 to 113 knots
(D) 114 to 135 knots
(E) >135 knots

68. An open access system, where there are no rules regulating resources, is described by

(A) externalizing costs
(B) "Tragedy of the Commons"
(C) the Clean Air Act
(D) the second law of thermodynamics
(E) the genuine progress index

69. All the following characteristics describe *r*-adapted species except

(A) long life
(B) rapid growth
(C) adaptation to varied environment
(D) niche generalist
(E) prey

70. If a population's growth is slow with low birth rates and low death rates, which demographic type would it describe?

(A) Preindustrial
(B) Industrial
(C) Transitional
(D) Postindustrial
(E) Pretransitional

71. When high levels of phosphates cause the overgrowth of aquatic plants and algae, high dissolved oxygen consumption, and fish death, it is called

(A) flocculation
(B) calcification
(C) aeration
(D) fossilization
(E) eutrophication

72. Which of the following is a disinfectant used in many water treatment plants?

(A) Toluene
(B) Arsenic
(C) Vinegar
(D) Chlorine
(E) Ammonia

73. An earthquake that moves heavy furniture and is felt by everyone but has slight overall damage is measured on the Richter scale (refer to Table 5.2) as

(A) 3.0 to 3.9
(B) 4.0 to 4.9
(C) 5.0 to 5.9
(D) 6.0 to 6.9
(E) 7.0 and higher

74. Which of the following is the main energy or heat source for the ocean's food chain?

(A) Magnesium nodules
(B) Hydroelectric power
(C) Sunlight
(D) Wind
(E) Bioluminescence

75. The categories, extremely arid, arid, and semiarid, describe

(A) deserts
(B) oceans
(C) tornadoes
(D) glaciers
(E) forests

76. The Intergovernmental Panel on Climate Change (IPCC) gathered over 200 scientists and national experts to work together to help countries create

(A) crop rotation schedules
(B) guidelines to atmospheric inventories across international borders
(C) World Bank lending methods
(D) green certificates
(E) a communal resource management system

77. Residence time is the

(A) time you spend at home after school
(B) emigration time of non-native species
(C) growth time of kelp beds
(D) time magma spends in the mantle
(E) time water spends in the groundwater part of the hydrologic cycle

78. The replacement birth rate is always

(A) 2
(B) the number an industrial country allows each couple
(C) one-half the number of children a couple has
(D) the number of children a couple needs to replace themselves in a population
(E) the number of grandchildren divided by the number of parents

79. What weather process has become the "bad boy" of the world's weather shifts?

(A) Evapotranspiration
(B) Trade winds
(C) El Niño–Southern Oscillation
(D) High humidity
(E) Glaciation

80. Aquifers that lie beneath layers of impermeable clay are known as

(A) unconfined aquifers
(B) confined aquifers
(C) dry aquifers
(D) fractured aquifers
(E) balanced aquifers

81. Ozone has a distinctive odor and is

(A) purple
(B) black
(C) blue
(D) yellow
(E) red

82. The bioprocess of taking DNA from one species and splicing it into the genetic material of another is known as

(A) genetic engineering
(B) biological sustainability
(C) genetic preservation
(D) evolution
(E) biodiversity

83. Pathogenic microorganisms cause

(A) weight gain
(B) increased water salinity
(C) diabetes
(D) typhoid fever
(E) reduced tourism

84. Globally, at least 1,000 insect species and 550 weed and plant pathogens have developed chemical resistance to

(A) pesticides
(B) sodium
(C) pollen
(D) rodenticides
(E) carbon

85. Which of the following is not known to cause water turbidity?

(A) Plankton
(B) River rocks
(C) Silt
(D) Clay
(E) Organic matter

86. Cost-effective photovoltaic systems are great for remote cabins, and beach or vacation homes without

(A) flashlight batteries
(B) sewer systems
(C) access to an electricity grid
(D) telephones
(E) landscape lighting

87. Water covers roughly what percent of Earth's surface?

(A) 27%
(B) 45%
(C) 64%
(D) 72%
(E) 80%

88. Evaporation occurs when

(A) water freezes
(B) water changes from a liquid to a gas
(C) plants turn toward the sun
(D) rainfall leads to runoff
(E) water resides underground for a long time

89. The amount of surface area needed to meet a population's needs and to dispose of its waste is its

(A) urban transit area
(B) population baseline
(C) ecological footprint
(D) public to federal land ratio
(E) human impact

90. A buildup of toxins in the tissues of predatory birds eating poisoned rodents and other small animals is an example of

(A) natural selection
(B) conservation of species
(C) biocide
(D) bioaccumulation
(E) lipid detoxification

91. The boundary between two watersheds is called a

(A) divide
(B) continental shelf
(C) water table
(D) scarp
(E) plateau

92. The two main ways geologists describe rock textures are grain size and

(A) smoothness
(B) size
(C) brittleness
(D) radioactivity
(E) color

93. Total organic carbon is used by hydrologists to

(A) determine pesticide uses
(B) study the crystalline structure of minerals
(C) check the health of freshwater
(D) classify sedimentary particles
(E) standardize no. 2 pencils

94. Circulating air and water patterns affected by the Earth's rotation are known as the

(A) geomorphic cycle
(B) carbonization effect
(C) geosynclinal cycle
(D) Coriolis effect
(E) El Niño–Southern Oscillation

95. Referring to Table 12.1, mercury and arsenic are in which class of pesticides?

(A) Natural organic
(B) Inorganic
(C) Microbial
(D) Organophosphates
(E) Carbamate

96. The rationale for protecting wilderness areas supports all the following except

(A) wildlife refuges and reproductive areas
(B) research basis for species' changes
(C) place of solitude and primitive recreation
(D) area of undisturbed natural beauty for future generations
(E) increased water pollution

97. Fecal coliform bacteria are naturally found in

(A) frozen ponds
(B) eggplants
(C) petrified wood
(D) feces and intestinal tracts of humans
(E) hot water tanks

98. Which of the following factors is not used to classify deserts?

(A) Volcanism
(B) Humidity
(C) Wind
(D) Total rainfall
(E) Temperature

99. Which greenhouse gas is about 145% higher now than it was in the 1800s?

(A) Butane
(B) Nitrogen
(C) Methane
(D) Oxygen
(E) Carbon dioxide

100. The three main rock types are sedimentary, metamorphic, and

(A) glacial
(B) crystalline
(C) freeze-fractured
(D) igneous
(E) opaque

Sample Exam 2—Free-Response Questions

1. In July of 1988, Yellowstone's wilderness was devastated by its worst forest fires since the park was founded in 1872. Weather conditions played a major factor in the fires spreading through all stages of Yellowstone's forests, as unseasonably high winds and drought took the place of summer rains that would have naturally impeded the fires' progress. The park's controversial "natural burn" policy established in 1968 allows for fires to run their natural course unless started by humans. From 1972 to 1987 the average area burned was 910 hectares. In 1988, an area of 321,273 hectares burned. Outcry from citizens and media fueled public fears over the mass devastation caused to the park, calling for its official to be fired and the "let it burn" policy to be changed.
   (a) Define succession, and compare primary and secondary succession. Is this an example of primary or secondary succession?
   (b) Why are forest fires a necessary part of succession?
   (c) What was the percent increase of area (in hectares) burned in 1988 compared to the average area burned from 1972 to 1987?
   (d) Why was (and is) the "natural burn" policy controversial, and what could have been done to allay the public's concerns about the policy?

2. Soaring gas prices, coupled with an economic down turn, have made many reconsider their vehicle of choice. During this time, the benefits of electric vehicles (EVs) became clear to the public. In urban areas, where air pollution is a huge environmental and health concern, EVs' zero emissions were a big plus. However, a study conducted by *Transport Watch*, a nongovernmental group in the UK, concluded that EVs are less efficient than diesel-powered engines and produce twice the CO_2 emissions in the UK than do diesel engines (i.e., when comparing the fossil fuels used to produce each vehicle type). When the fuel lost (i.e., point source → fueling station → tank → wheels) for each is measured, only 24% of the energy produced at the point source reaches EVs for use, while 45% reaches diesel engines.
   (a) How do CO_2 emissions affect air quality?
   (b) For countries depending primarily on coal, like China, what could a switch to electric vehicles mean for the environment?
   (c) Describe two measures to raise or offset electric vehicle efficiency.
   (d) Describe two ways energy can be lost along the supply chain.

3. Atlanta, Georgia, lies near the headwaters of the Chattahoochee River. The Chattahoochee has been one of the most polluted rivers in the United States for decades. Over the last century Atlanta and other cities along the "Hooch" ran sewage directly into the river. Continued industrial, municipal, and agricultural waste in the river keeps levels of fecal coliform bacteria at or near amounts dangerous to humans and other wildlife. Along with pollution concerns, the environmental impact of human civilization along the Chattahoochee has had many adverse effects on local flora and fauna, leading many species relying on its waters to become endangered or extinct.
   (a) Describe two reasons for increased levels of fecal coliform levels in the river.
   (b) Fecal coliform levels are measured using indicator organisms. What are indicator organisms and how do they help measure the health of an ecosystem?
   (c) Identify one point source and one nonpoint source example of pollution harming the Chattahoochee River.
   (d) Of what significance, if any, is Atlanta's geographical location near the headwaters of the Chattahoochee?

4. A recent study published in the *New England Journal of Medicine* gives statistical validity to the common wisdom, "Clear skies equal healthy people." Taking data from 51 cities across the United States from 1979 to 2000, the authors found average life spans in the subject areas rose an average of three years, from 74 to 77. Cleaner air was a factor linked to increased human longevity. Even tiny air pollution amounts can make a difference to health. Cleaning up just 10 micrograms of pollutant per meter of air could add nine months to an average life span. (Source, *Time Magazine,* January 22, 2009.)

(a) Describe three human-made means by which air is contaminated in urban environments.

(b) Describe three contaminants that, in excess, pollute the Earth's air.

(c) Which policies has the United States enacted to reduce air pollution?

(d) Describe three ways in which air pollution can be further reduced.

› Sample Exam 1—Answers

1. **D**—The number of open spaces or pores between soil particles give it porosity.
2. **C**—These greenhouse gases add to the enhanced greenhouse effect.
3. **E**—Water is stored in reservoirs for varying amounts of time.
4. **E**—CAFE Tier 2 standards reduced nitrous oxide and sulfur emissions in vehicles.
5. **D**—Greenhouse gases are naturally balanced in the environment, but after industrialization, they have risen dramatically.
6. **A**—Neutral pH is 7.0 with acidic pH at <7.0 and basic pH at >7.0.
7. **D**
8. **A**—Carbon dioxide is produced by burning of fossil fuels and from natural sources.
9. **C**
10. **E**—Heat from the core moves upward to the surface through volcanism.
11. **E**
12. **B**—Water is pulled up from the ground through the roots and finally released through pores in the leaves.
13. **C**—They are all associated with living, biological processes.
14. **D**
15. **E**—Iceland's turn from burning fossil fuels to geothermal energy has reduced pollution considerably.
16. **D**—Carson's book was the first to link uncontrolled pesticide and chemical use to environment impacts.
17. **B**
18. **C**
19. **E**
20. **D**—Waste is labeled hazardous specifically for its toxicity.
21. **A**—The movie described the potential impacts of nuclear core meltdown.
22. **B**—The oldest rock layers are below newer rock unless plate tectonics have shifted the layers.
23. **C**—Infiltration is the entry of water into the soil and between rock particles.
24. **E**—Wind and windmills were an early and simple power producer.
25. **B**—Low-flow toilets, which use less water, are required in several states.
26. **C**
27. **A**—Geological plates are part of the plate tectonic process.
28. **E**—Fossil fuel burning creates a variety of greenhouse gases.
29. **D**
30. **B**—Runoff often carries pesticides and fertilizers into waterways as well.
31. **A**—Carbon is the foundational element of all organic substances and organisms.
32. **D**—France lessened its need for foreign oil by investing in nuclear energy.
33. **C**—Nonpoint source pollutants are not from a single source, but many combined.
34. **E**—The mesosphere stretches 80–90 km and lies above the stratosphere and below the thermosphere.
35. **A**—Ocean and plant processes hold carbon in their structures until released later by burning or decay.
36. **D**—The jet stream has a significant impact on weather patterns.
37. **B**—Methane is produced by many sources and adds up in the atmosphere.
38. **B**—There are three types of hydroelectric power (also known as hydropower) plants: impoundment, diversion, and pumped storage.

39. **C**—Salmon are not known as water contaminants.

40. **E**

41. **C**—Most commercial air travel takes place in the lower part of the stratosphere.

42. **E**—When chemical bonds are broken in a radioactive element and energy is given off, the element is said to decay.

43. **C**

44. **A**—Whale oil was used before kerosene, but was not given as an answer choice.

45. **C**—As water moves through the hydrology cycle, pollutants enter it at specific points or from nonpoint sources as part of runoff or from common farming sources.

46. **D**—The photovoltaic reaction produces voltage when exposed to radiant (light) energy.

47. **B**—Plants and trees use carbon in their physiological processes.

48. **C**

49. **A**—Dendritic drainage has a main stream that splits to smaller and smaller streams.

50. **D**

51. **B**—Turbines catch the wind with two to three propeller-like blades mounted on a shaft to form a rotor that spins a generator creating electricity.

52. **D**—Solar flares hit the magnetosphere and pull electrons from atoms, which causes magnetic storms near the poles. Red and green atmospheric lights are seen when scattered electrons reunite with atoms.

53. **C**—Sometimes oil seeps back into an old site, and other times new methods make retrieval possible.

54. **E**

55. **B**—Uranium is mined in pit mines around the world. Miners have to be careful about bringing radioactive dust home on their clothes and boots.

56. **C**—About this time the public realized fossil fuels (e.g., oil) were finite.

57. **C**—The Continental Divide in the Rocky Mountains diverts water east and west. Western water drains to the Pacific.

58. **B**—Sunlight entering a greenhouse is trapped and heats its interior.

59. **D**—In 2014, 1,199 (22%) of mammal species were designated as vulnerable or endangered. Another 188 mammal species (3%) were reported as critically endangered.

60. **C**

61. **B**—Wind removes the insulating layer of air above the skin and chills the body.

62. **D**—The water table marks the boundary between the zone of saturation and zone of aeration.

63. **D**—Higher-order streams have increasingly larger areas of drainage and flow rates.

64. **E**

65. **C**—Thermosphere temperatures are affected by high or low sunspot and solar flare activity. The greater the sun's activity, the hotter the thermosphere.

66. **D**

67. **D**

68. **B**—Garret Hardin wrote, "The Tragedy of the Commons," an article explaining how commonly held resources are often misused since no one person or group claims responsibility.

69. **A**—*r*-adapted species have short lives since they are usually prey.

70. **B**—The industrial demographic type in developing countries has low birth rates and low death rates.

71. **E**—When sediments enter a lake, fish respiration is impacted, plant growth and water depth are limited, and aquatic organisms asphyxiate.

72. **D**—Chlorine stops microbial problems before they start and has lasting effects.

73. **C**

74. **C**—Sunlight is a practically inexhaustible power source for Earth.

75. **A**—Most deserts are categorized with regard to some level of aridity depending on rainfall.

76. **B**—Policy makers and regulatory agencies use these inventories to check policy compliance and emission rates.

77. **E**—Different global reservoirs hold water during groundwater residence times.

78. **D**—Statistically the human replacement birth rate can be as high as 2.5.

79. **C**—Predicting this cyclical weather pattern in the Pacific, Atlantic, and Indian Oceans is important since ENSO events have global impacts.

80. **B**—Water is blocked by impermeable clay layers of a confined aquifer.

81. **C**

82. **A**—Genetic engineering is the process used to change the genetic material of an organism to produce desired traits.

83. **D**

84. **A**—Pesticide resistance is a problem since every target species has genetically diverse individuals who are resistant to the pesticide.

85. **B**—All the choices, except river rocks, cause a degree of turbidity.

86. **C**—They provide self-contained power or backup power off the grid.

87. **D**

88. **B**—Evaporation is the process where water moves from the liquid to the gaseous phase.

89. **C**—Ecological footprint measures human demand on the environment.

90. **D**—Bioaccumulation is also found in the tissues of lower organisms like zooplankton.

91. **A**—The Continental Divide in North America is the high line running through the Rocky Mountains.

92. **E**—Grain size and color are the two main ways geologists distinguish rock texture.

93. **C**—Organic content is commonly measured as total organic carbon and dissolved organic carbon, which are essential components of the carbon cycle.

94. **D**—The Coriolis effect is a perceived deflection of moving objects viewed from a turning frame of reference. An optical illusion in terms of north-south rotating motion, air moving from the north pole seems to turn right (northern hemisphere) and left (southern hemisphere) due to the Earth's rotation.

95. **B**—Mercury and arsenic are inorganic elements.

96. **E**—Protected pristine areas are meant to be protected from all types of pollution.

97. **D**—Microorganisms like viruses and bacteria cause disease when not eliminated at water treatment plants.

98. **A**—Volcanism is not considered a current factor in desertification.

99. **C**

100. **D**—The three main rock types are sedimentary, metamorphic, and igneous.

› Sample Exam 2—Answers

1.

a. Succession describes the rise and establishment of natural communities and their replacement over time. Two general types of succession are primary and secondary. Primary succession takes place on an area previously devoid of life, like the aftermath of a lava flow. Secondary succession takes place in an area where life exists but has been largely destroyed, as with a forest fire.

b. Fires are a natural event in the life of a healthy forest. Sometimes, fires are essential for a species' regeneration (e.g., red pines). Generally, fires move through the understory of forests burning dead and fallen wood and lower branches. This helps forest regeneration by providing ideal growing conditions for both plants and animals while also giving them needed space to grow. Fires also help stem the spread of disease or blight in forests.

c. There was a 353% increase in burn area.

d. The "natural burn" policy was controversial largely due to public ignorance on the benefits of natural and controlled forest fires. Lack of public knowledge in 1988, along with media coverage, added to the extreme Yellowstone case and made the policy even more controversial at the time. To help calm public concerns over park policy, more could have been done to educate the public on the benefits of wildfires in forests. Ad campaigns and school programs would have helped better inform citizens and decrease the panic following the 1988 fire season.

2.

a. Carbon dioxide (CO_2) emissions are harmful for atmospheric health because of the large quantities produced by humans. The atmosphere naturally stores carbon dioxide as part of the carbon cycle, but elevated CO_2 levels have been shown to lead to sea level changes, snow melt, disease, and heat stress. Because it is a greenhouse gas, high concentrations of atmospheric CO_2 lead to global warming through raising the atmosphere's ability to absorb and emit thermal infrared radiation.

b. If the EV efficiency numbers are correct, for countries like China, which rely on coal-fired plants to generate electricity, switching to electric vehicles could dramatically raise CO_2 emissions and have drastic effects on public health and the environment.

c. Creating cleaner energy sources to produce electricity could help raise the efficiency level. To raise the efficiency level, methods that waste or lose less "fuel" from point source to tank should be used.

d. Along the supply chain, energy can be lost through transporting electricity to the "fueling" stations, and electricity can be lost at the power station where the car is plugged in.

3.

a. Municipal and industrial waste (e.g., sewage) released directly into the river presents one way fecal coliform levels could rise. Water level is another factor. Less sewage from industry or agricultural runoff could have a similar effect as bacteria levels if the water level is low due to drought.

b. Indicator organisms are microbes whose presence in water signals the presence of fecal material and other contaminants. Because they are easy to measure, the number of organisms in a water sample can be measured against the amount in a normal water supply to measure the water's health.

c. An example of a polluting river point source would be a drainage or sewage pipe spilling contaminants directly into the river. A nonpoint source example would be runoff from an agricultural source due to flooding or heavy rains.
d. Atlanta is a large metropolis and has tons of waste. Its location at the river's headwaters means that its pollutants affect everything else downstream, increasing its harmful environmental impact.

4.

a. Plants burning coal, unless otherwise outfitted for prevention, emit CO_2 into the air. The cars we drive further emit CO_2. Recreational and industrial vehicles like snowmobiles and tractors also emit CO_2.
b. Carbon dioxide, sulfur dioxide, carbon monoxide.
c. 1955 Air Pollution Control Act, 1963 Clean Air Act, 1967 Air Quality Act, 1970 Clean Air Extension, 1990 Clean Air Act.
d. Clean energy sources like solar and wind power can be used to reduce air pollution. Better means of transportation (e.g., van pools, bicycles, and public transportation) can also benefit air quality especially in urban environments. Clean coal technology can be put into practice, especially in countries heavily reliant on coal as fuel, to greatly reduce the atmospheric contaminants released from coal power plants.

Appendixes

UNITS AND CONVERSION FACTORS

Standard International (SI) Units

PREFIX	ABBREVIATON	MULTIPLES OF TEN	NUMBER	NAME
zetta	Z	10^{21}	1,000,000,000,000,000,000,000	sextillion
exa	E	10^{18}	1,000,000,000,000,000,000	quintillion
peta	P	10^{15}	1,000,000,000,000,000	quadrillion
tera	T	10^{12}	1,000,000,000,000	trillion
giga	G	10^{9}	1,000,000,000	billion
mega	M	10^{6}	1,000,000	million
kilo	k	10^{3}	1,000	thousand
hecto	h	10^{2}	100	hundred
deka	da	10^{1}	10	ten
deci	d	10^{-1}	1/10	tenth
centi	c	10^{-2}	1/100	hundredth
milli	m	10^{-3}	1/1,000	thousandth
micro	µ	10^{-6}	1/1,000,000	millionth
nano	n	10^{-9}	1/1,000,000,000	billionth
pico	p	10^{-12}	1/1,000,000,000,000	trillionth
femto	f	10^{-15}	1/1,000,000,000,000,000	quadrillionth
atto	a	10^{-18}	1/1,000,000,000,000,000,000	quintillionth
zepto	z	10^{-21}	1/1,000,000,000,000,000,000,000	sextillionth

SI Units and SI-English Unit Conversions

Length

The SI base unit for length is the *meter* (m).

1 centimeter (cm)	0.3937 inch
1 inch (in)	2.5400 centimeters
1 meter (m)	3.2808 feet; 1.0936 yards
1 foot (ft)	0.3048 meter
1 yard (yd)	0.9144 meter
1 kilometer (km)	0.6214 mile (statute); 3,281 feet
1 mile (mi) (statute)	1.6093 kilometers
1 mile (nautical)	1.8531 kilometers
1 fathom	6 feet; 1.8288 meters
1 angstrom (Å)	10^{-8} centimeters
1 micrometer (µm)	0.0001 centimeters

Velocity

The SI base unit for velocity is the *kilometer/hour* (km/h).

1 kilometer (km)/hour	27.78 centimeters/second (cm/s)
1 mile (mi)/hour	17.60 inches/second

Area

The SI base unit for area is the *square meter* (m^2).

1 square centimeter	0.1550 square inch
1 square inch	6.452 square centimeters
1 square meter	10.764 square feet; 1.1960 square yards
1 square foot	0.0929 square meter
1 square kilometer	0.3861 square mile
1 square mile	2.590 square kilometers
1 acre (U.S.)	4,840 square yards

Volume

The SI base unit for volume is the *cubic meter* (m^3).

1 cubic centimeter	0.0610 cubic inch
1 cubic inch	16.3872 cubic centimeters
1 cubic meter	35.314 cubic feet
1 cubic foot	0.02832 cubic meter
1 cubic meter	1.3079 cubic yards
1 cubic yard	0.7646 cubic meter
1 liter	1,000 cubic centimeters
1 gallon (U.S.)	3.7853 liters

Temperature

The SI base unit for temperature is the *kelvin* (K).

Celsius to Fahrenheit:	$°F = \left(\frac{9}{5}\right) °C + 32$
Fahrenheit to Celsius:	$°C = \left(\frac{5}{9}\right) °F - 32$
Celsius to kelvin:	$K = °C + 273.15$

Mass

The SI base unit for mass is the *kilogram* (kg).

1 gram	0.03527 ounce
1 ounce	28.3496 grams
1 kilogram	2.20462 pounds
1 pound (lb)	0.45359 kilogram (kg)
1 metric ton (t)	10^3 kg

Pressure

The SI base unit for pressure is the *pascal* (Pa).

1 millimeter of mercury (mmHg)	1 torr
1 pascal (Pa)	1 $N/m^2 = 1\ kg/m \cdot s^2$
1 atmosphere (atm)	1.01325 10^3 Pa = 760 torr
1 bar	1 × 10^5 Pa

Energy

The SI base unit for energy is the *joule* (J).

1 joule (J) = force exerted by a current of 1 amp/second passing through a resistance of 1 ohm

1 watt (W)	1 joule/second
1 kilowatt-hour (kWh)	1,000 (10^3) watts exerted for 1 hour
1 megawatt (MW)	1 million (10^6) watts
1 joule	1 kg m^2/s^2 = 1 coulomb volt
1 calorie (cal)	4.184 joules
1 food Calorie (Cal)	1 kilocalorie (kcal) = 4,184 joules
1 British thermal unit (BTU)	252 calories (cal) = 1,053 joules
1 standard barrel (bbl) of oil	42 gallons (160 liters) or 5.8 BTUs
1 metric tone of standard coal	27.8 million BTUs or 4.8 bbl oil

ACRONYMS

ACRONYM	DESCRIPTION
AA	Attainment area
ACL	Alternative concentration limit
AMS	American Meteorological Society
AO	Agent Orange—an herbicide and defoliant containing dioxin used in Vietnam by the United States
API	American Petroleum Institute
AQCR	Air quality control region
ARAR	Applicable relevant and appropriate requirements (cleanup standards)
ASTM	American Society for Testing and Materials
ATSDR	Agency for Toxic Substances and Disease Registry
AWEA	American Wind Energy Association
BACM	Best available control measure
BACT	Best available control technology
BAT	Best available technology
BATEA	Best available technology economically achievable
BCT	Best conventional technology
BDAT	Best demonstrational achievable (also "available") technology
BDT	Best demonstrational technology
BEJ	Best engineering judgment
BIF	Boiler and industrial furnace
BMP	Best management practice
BOD	Biological oxygen demand—index of amount of oxygen used by bacteria to decompose organic waste
BPJ	Best professional judgment
BPT	Best practical control technology
BRS	Biennial reporting system
BTU	British thermal unit—energy required to raise 1 lb of water 1°F
C&D	Construction and demolition
CAA	Clean Air Act
CAIR	Comprehensive assessment information rule
CAMU	Corrective action management unit
CAS	Chemical abstract service
CBI	Confidential business information
CCP	Commercial chemical product
CDD	Chlorodibenzodioxin
CDF	Chlorodibenzofuran
CEM	Continuous emission monitoring
CEQ	Council on environmental quality
CERCLA	Comprehensive Environmental Response, Compensation, & Liability Act of 1980 (amended 1984)
CESQG	Conditionally exempt small quantity generator of hazardous wastes
CFC	Chlorofluorocarbon—an ozone-depleting refrigerant

(*Continued*)

ACRONYM	DESCRIPTION
CFR	Code of Federal Regulations
CHEMTREC	Chemical Transportation Emergency Center
CHIPS	Chemical Hazards Information Profiles
CIESIN	Center for International Earth Science Information Network
CNG	Compressed natural gas
COD	Chemical oxygen demand
CPSC	Consumer Product Safety Commission (16 CFR)
CTG	Control techniques guidelines
CWA	Clean Water Act
DCO	Delayed compliance order
DDT	Dichlorodiphenyltrichloroethane—a toxic pesticide
DMR	Discharge monitoring reports
DNA	Deoxyribonucleic acid—made of phosphates, sugars, purines, and pyrimidines; helix shape carries genetic information in cell nuclei
DO	Dissolved oxygen
DOD	Department of Defense
DOE	Department of Energy
DOJ	Department of Justice
DOT	Department of Transportation
DRE	Destruction and removal efficiency
EERE	Energy efficiency and renewable energy
EIS	Environmental impact statement
ELF	Extremely low frequency electromagnetic wave (< 300 Hz)—emitted by electrical power lines
EMS	Environmental Management System (also see ISO 14000)
EP	Extraction procedure
EPA	Environmental Protection Agency
EPR	Extended product responsibility
EREF	Environmental Research and Education Foundation
ERT	Environmental Resources Trust Inc.
ESA	Environmental site assessment
ESI	Environmental Sustainability Index
ESP	Electrostatic precipitator
EWEA	European Wind Energy Association
FDA	Food and Drug Administration (21 CFR)
FFCA	Federal Facility Compliance Act
FIFRA	Federal Insecticide, Fungicide and Rodenticide Act
First Third	August 17, 1988, Federal Register (53 FR 31138)—the first of the hazardous waste land disposal restrictions
FR	Federal Register
GATT	General Agreement on Tariffs and Trade—since 1947; over 100 member countries
GCM	Global Climate Model
GLP	Good laboratory practices
GMER	Green Mountain Energy Resources
GMO	Genetically modified organism
GMP	Good manufacturing procedures

(*Continued*)

ACRONYM	DESCRIPTION
GPO	Government printing office
GRAS	Generally recognized as safe
HazMat	Hazardous material
HAZWOPER	29 CFR 1910.120—the OSHA/EPA requirement to have all employees trained if they will be handling, managing, or shipping hazardous wastes
HHW	Household hazardous waste
HRS	Hazard ranking system
HSWA	Hazardous and Solid Waste Amendments—1984
HW	Hazardous waste
HWM	Hazardous waste management
Hz	Hertz—frequency with which alternating current changes direction
ID	Hazardous waste identification number assigned to RCRA generators, transporters, and TSDFs
IEA	International Energy Association
INPO	Institute of Nuclear Power Operations
INUR	Inventory update rule
IPM	Integrated pest management
ISB	In situ burning
ITC	Interagency Testing Committee
kWh	Kilowatt-hour
LAER	Lowest achievable emission rate
LCA	Life cycle analysis/assessment
LDR	Land disposal restrictions (40 CFR Part 268)
LED	Light-emitting diode
LEPC	Local emergency planning committee
LNG	Liquid natural gas
LOEL	Lowest observed effect level
LPG	Liquid petroleum gas (or propane)
LQG	Large quantity generator of hazardous wastes—this term has a specific definition under RCRA
LUST	Leaking underground storage tanks
MACT	Maximum achievable control technology
Maglev	Magnetic levitation train—using magnetic forces for high-speed movement
MARPOL	International Convention on the Prevention of Pollution from Ships
MCL	Maximum concentration limit or level
MRQ	Monthly Hotline Report Q&A—the Hotline prepares a monthly report that contains questions and answers on common or difficult RCRA topics. The EPA publishes this report. The questions and answers can be *usually* used as EPA guidance. Beware: The Hotline is run by an EPA contractor whose answers are often erratic and *not* legally binding.
MRT	Mean residence time—the amount of time a water molecule spends in a reservoir before moving on
MSDS	Material safety data sheet (under OSHA)
MSW	Municipal solid waste (trash and nonhazardous waste)

(*Continued*)

ACRONYM	DESCRIPTION
MW	Megawatt—1,000 kilowatts (1 million watts)
NAA	Nonattainment area
NAAQS	National Ambient Air Quality Standards
NAEWG	North American Energy Working Group
NAFTA	North American Free Trade Agreement—Canada, Mexico, and the United States
NCAR	National Center for Atmospheric Research
NCP	National contingency plan
NEPA	National Environmental Policy Act
NESHAP	National Emissions Standard for Hazardous Air Pollutants
NGO	Nongovernmental organizations—over 10,000 organizations worldwide linked by ECONET
NIOSH	National Institute of Occupational Safety and Health
NIPDWR	National Interim Primary Drinking Water Regulation
NPDES	National Pollutants Discharge Elimination System
NPL	National Priorities List—list of Superfund sites
NRC	Nuclear Regulatory Commission
NSDWR	National Secondary Drinking Water Regulation
NSPS	New Source Performance Standards
NSR	New Source Review
NTIS	National Technical Information Service (usually charges you for the same information that search engines provide for free)
ODC	Ozone depleting chemical
OGF	Old-growth forest—high-biodiversity ecosystem with trees 300 to 1,000 years old
OH	Hydroxyl radicals—atmospheric molecule that reduces methane (CH_4), carbon monoxide (CO), and ozone
OPM	Operation and maintenance
ORD	Office of Research and Development
OSHA	Occupational Safety and Health Administration (29 CFR)
OSW	Office of Solid Waste
OSWER	Office of Solid Waste and Emergency Response
OTG	Off-the-grid power generation independent of a major power plant
PAIR	Preliminary Assessment Information Rule
PCB	Polychlorinated biphenyl—used in dyes, paints, light bulbs, transformers, and capacitors
PEL	Permissible exposure limit
pH	Logarithmic scale that measures acidity (pH 0) and alkalinity (pH 14); pH 7 is neutral
PM-10	Particulate matter < 10 micrometers
PNA	Polynuclear aromatic compounds
PNIN	Premanufacture notification
POP	Publicly owned treatment works
PPB	Parts per billion
PPM	Parts per million
PRP	Potentially responsible parties
PSD	Prevention of significant deterioration

(*Continued*)

ACRONYM	DESCRIPTION
PSP	Point source pollution
PV	Photovoltaic device—generates electricity through semiconducting material
PVC	Petrochemical formed from toxic gas vinyl chloride and used as a base in plastics
QA/QC	Quality assurance/quality control
R	Richter scale—logarithmic scale (0–9) used to measure the strength of an earthquake
Rad	Radiation absorbed dose—amount of radiation energy absorbed in 1 gram of human tissue
R&D	Research and development
RACM	Reasonably available control measure
RACT	Reasonably available control technique
RCRA	Resource Conservation and Recovery Act of 1976—resulted in hazardous waste regulations
rDNA	Recombinant DNA—new mix of genes spliced together on a DNA strand; (biotechnology)
Rem	Roentgen (R) equivalent man—biological effect of a given radiation at sea level is 1 rem.
RI/FS	Remedial Investigation/Feasibility Study
RNA	Ribonucleic acid—formed on DNA and involved in protein synthesis
RPCC	Release prevention, control, and countermeasure
RQ	Reportable quantity
RUST	RCRA underground storage tanks
SARA	Superfund Amendments and Reauthorization Act
SBS	Sick-building syndrome
S&B A	Slash & burn agriculture
SDWA	Safe Drinking Water Act of 1974
Second Third	June 23, 1989, Federal Register (54 FR 26594)—hazardous waste land disposal restrictions
SERC	State Emergency Response Commission
SIC	Standard Industrial Classification
SIP	State Implementation Plan
SMCL	Secondary maximum contamination level
SOC	Schedule of compliance
SPCC	Spill prevention control and countermeasures
SPDES	State pollutant discharge elimination
SQG	Small quantity generator of hazardous wastes (has a specific definition)
SW-846	Test methods for evaluating solid waste, physical/chemical methods
SWMU	Solid waste management unit
TCDD	Tetrachlorodibenzodioxin
TCE	Tetrachloroethylene, perchloroethylene
TCLP	Toxic characteristic leaching procedure
Third Third	June 1, 1990, Federal Register (55 FR 22520)—refers to hazardous wastes "landban" land disposal restrictions
TOC	Total organic carbon
TRU	Transuranic wastes

(*Continued*)

ACRONYM	DESCRIPTION
TSCA	Toxic Substances Control Act of 1976—regulates asbestos, PCBs, new chemicals being developed for sale, and other chemicals
TSDF	Treatment, storage, or disposal facility (permitted hazardous waste facility)
TSS	Total suspended solids
TWA	Time weighted average
TWC	Third-world countries
UIC	Underground injection control
USCG	United States Coast Guard
USDA	United States Department of Agriculture
USDW	Underground source of drinking water
USGS	United States Geological Survey—manages LandSat which images the environment via satellite
USPS	United States Postal Service
UV	Ultraviolet radiation from the sun (UVA, UVB types)
VOC	Volatile organic compound—carbon-containing compounds that evaporate easily at low temperatures
VOME	Vegetable oil methyl ester—biodiesel derived from the reaction of methanol with vegetable oil
W	Watt—unit of electrical power
WAP	Waste analysis plan
WB	World Bank—owned by governments of 160 countries and funds hydroelectric plants and encourages ecotourism; does not fund nuclear energy
WCU	World Conservation Union
WGI	World Glacier Inventory
WHS	World Heritage Site—natural or cultural site recognized as globally important and deserving international protection
WRI	World Resources Institute

BIBLIOGRAPHY

Ball, P., 2000, *Life's Matrix: A Biography of Water*, Farrar, Straus and Giroux, New York, NY.

Braimoh, A. K., and Vlek, P.L.G., editors, 2008, *Land Use and Soil Resources*, Springer Science+Business Media, NY, NY.

Cunningham, W. and Cunningham, M., 2006, *Principles of Environmental Science*, McGraw Hill Companies, Inc., New York, NY.

Deffeyes, K. S., 2001, *Hubbert's Peak: The Impending World Oil Shortage*, Princeton University Press, Princeton, NJ.

Department of Energy, 2008, *Renewable Energy Trends in Consumption and Electricity 2006*, Energy Information Administration, Office of Coal, Nuclear, Electric, and Alternate Fuels, Washington, DC.

Energy Information Administration, Office of Integrated Analysis and Forecasting, *International Energy Outlook 2004*, rep. no. DOE/EIA-0484(2004), U.S. Department of Energy, Washington, DC.

Environmental Protection Agency, 2002, *In Brief: The U.S. Greenhouse Gas Inventory*, Office of Air and Radiation (EPA 430-F-02-008), Washington, DC.

Gallant, R. A., 2003, *Structure: Exploring Earth's Interior*, Benchmark Books, New York, NY.

Goodstein, D., 2004, *Out of Gas: The End of the Age of Oil*, W.W. Norton & Company, New York, NY.

Lambert, D., 2006, *The Field Guide to Geology*, Facts on File, Inc., 2nd Edition, Diagram Visual Information, Inc., New York, NY.

Lovelock, J., 2007, *The Revenge of Gaia: Earth's Climate Crisis and the Fate of Humanity*, Perseus Publishing, Cambridge, MA.

Lüsted, M. and Lüsted, G., 2005, *A Nuclear Power Plant*, Lucent Books, New York, NY.

Morris, N., 2007, *Global Warming*, World Almanac Library, Milwaukee, WI.

Mycle Schneider Consulting, 2008, *Bulletin of Atomic Scientists*, September 2008, pp 1–3.

Reynolds, R., 2005, *Guide to Weather*, Firefly Books, Ltd., Buffalo, NY.

Rogers, P., 2008, "Facing the Freshwater Crisis", *Scientific American*, August 2008, pp. 46–53.

Schacker, M., 2008, *A Spring without Bees: How Colony Collapse Disorder Has Endangered Our Food Supply*, Lyons Press, Guilford, CT.

Walker, G., 2007, *Why the Wind Blows and Other Mysteries of the Atmosphere*, Harcourt Books, Inc., Orlando, FL.

Williams, L. D. 2004, *Earth Science Demystified*, McGraw Hill Companies, Inc., New York, NY.

Williams, L. D. 2005, *Environmental Science Demystified*, New York, McGraw Hill Companies, Inc., New York, NY.

WEBSITES

Useful Environmental Science websites to use as resources in preparation for the AP Environmental Science exam.

Atmosphere

- http://epa.gov/climatechange/ghgemissions

Biodiversity

- http://www.cbd.int
- http://www.conservation.org/learn/biodiversity/Pages/overview.aspx

Earth's Formation

- http://hubble.nasa.gov
- http://www.ghcc.msfc.nasa.gov

Ecology

- http://www.itopf.com/information-services/data-and-statistics/statistics
- http://earthtrends.wri.org
- http://www.footprintnetwork.org
- http://www.sciencedaily.com/releases/2008/05/080521205303.htm
- http://www.fas.org/sgp/crs/misc/RL33938.pdf
- http://www.fas.org/sgp/crs/misc/RS21232.pdf
- http://www.blm.gov/wo/st/en/prog/grazing.html

Fossil Fuels

- http://www.evostc.state.ak.us/http://pubs.usgs.gov/dds/dds-060
- http://www.instituteforenergyresearch.org/energy-overview/fossil-fuels
- http://www.livescience.com/6215-oil-production-peak-2014-scientists-predict.html
- http://yearbook.enerdata.net
- http://www.fossil.energy.gov/education/energylessons/coal/MS_Coal_Studyguide_draft1.pdf

General

- http://geodata.grid.unep.ch
- http://apcentral.collegeboard.com/apc/public/repository/ap-environmental-science-course-description.pdf
- http://www.eia.gov/forecasts/ieo

Geothermal

- http://energy.sandia.gov/?page_id=381
- http://www.geothermal-energy.org

Glaciers

- https://nsidc.org/cryosphere/glaciers
- www.coolantarctica.com/toc.htm

Global Warming

- http://www.epa.gov/climatechange/index.html
- http://www.ipcc.ch

Green Energy

- http://www.winrock.org
- http://www.eere.energy.gov
- http://www.green-energy-news.com

Nanotechnology

- www.cnst.rice.edu
- www.nano.gov

Nuclear Energy

- http://www.epa.gov/cleanenergy/energy-and-you/affect/nuclear.html
- http://www.nei.org

Oceans/Water

- www.usgs.gov
- http://www.weather.gov
- http://www.marine-conservation.org
- http://water.epa.gov/polwaste/nps/index.cfm
- http://news.sciencemag.org/sciencenow/2004/07/15-01.html
- http://www.oceanenergycouncil.com/index.php/Tidal-Energy/Tidal-Energy.html

Solar

- http://www.solarbuzz.com/going-solar/understanding/technologies
- https://solarenergy.com/
- http://www.seia.org

Space

- www.nasa.gov/home/index.html
- http://nssdc.gsfc.nasa.gov/photo_gallery
- http://earth.jsc.nasa.gov/sseop/efs
- www.noaa.gov/satellites.html

Wastes

- www.epa.gov/osw/
- http://www.epa.gov/superfund/index.htm
- http://www.itopf.com/marine-spills/effects

Weather

- www.weather.com
- www.theweathernetwork.com

Wind

- www.awea.org
- http://www.renewableuk.com/
- http://www.windpoweringamerica.gov